国外优秀物理著作
原 版 系 列

电磁学——问题与解法（英文）

Electromagnetism—Problems and solutions

[美] 卡罗来纳·C. 伊利耶（Carolina C. Ilie）
[美] 撒迦利亚·S. 施雷森戈斯特（Zachariah S. Schrecengost）
著

哈尔滨工业大学出版社
HARBIN INSTITUTE OF TECHNOLOGY PRESS

黑版贸审字 08－2020－163 号

图书在版编目(CIP)数据

电磁学:问题与解法:英文/(美)卡罗来纳·C.伊利耶(Carolina C. Ilie),(美)撒迦利亚·S.施雷森戈斯特(Zachariah S. Schrecengost)著.—哈尔滨:哈尔滨工业大学出版社,2021.7

书名原文:Electromagnetism:Problems and solutions

ISBN 978－7－5603－9497－8

Ⅰ.①电… Ⅱ.①卡… ②撒… Ⅲ.①电磁学－高等学校－教材－英文 Ⅳ.①O441

中国版本图书馆 CIP 数据核字(2021)第 113671 号

Electromagnetism: Problems and Solutions Copyright © 2016 by Morgan & Claypool Publishers
All rights reserved.
The English reprint rights arranged through Rightol Media
(本书英文影印版权经由锐拓传媒取得 Email:copyright@rightol.com)

策划编辑	刘培杰 杜莹雪	
责任编辑	刘立娟 李兰静	
封面设计	孙茵艾	
出版发行	哈尔滨工业大学出版社	
社　　址	哈尔滨市南岗区复华四道街 10 号　邮编 150006	
传　　真	0451－86414749	
网　　址	http://hitpress.hit.edu.cn	
印　　刷	哈尔滨博奇印刷有限公司	
开　　本	720 mm×1 000 mm　1/16　印张 13　字数 210 千字	
版　　次	2021 年 7 月第 1 版　2021 年 7 月第 1 次印刷	
书　　号	ISBN 978－7－5603－9497－8	
定　　价	88.00 元	

(如因印装质量问题影响阅读,我社负责调换)

To my family, my mentors, and my students — CCI

To my friends, family, and mentors — ZSS

Contents

Preface	vi
Acknowledgements	viii
About the authors	ix

1   Mathematical techniques   1-1

1.1	Theory	1-1
	1.1.1 Dot and cross product	1-1
	1.1.2 Separation vector	1-1
	1.1.3 Transformation matrix	1-2
	1.1.4 Gradient	1-2
	1.1.5 Divergence	1-2
	1.1.6 Curl	1-3
	1.1.7 Laplacian	1-3
	1.1.8 Line integral	1-4
	1.1.9 Surface integral	1-4
	1.1.10 Volume integral	1-4
	1.1.11 Fundamental theorem for gradients	1-4
	1.1.12 Fundamental theorem for divergences (Gauss's theorem, Green's theorem, divergence theorem)	1-4
	1.1.13 Fundamental theorem for curls (Stoke's theorem, curl theorem)	1-4
	1.1.14 Cylindrical polar coordinates	1-4
	1.1.15 Spherical polar coordinates	1-5
	1.1.16 One-dimensional Dirac delta function	1-5
	1.1.17 Theory of vector fields	1-5
1.2	Problems and solutions	1-5
	Bibliography	1-34

2   Electrostatics   2-1

2.1	Theory	2-1
	2.1.1 Coulomb's law	2-1
	2.1.2 Electric field	2-1
	2.1.3 Gauss's law	2-2
	2.1.4 Curl of $\vec{E}$	2-2
	2.1.5 Energy of a point charge distribution	2-2

		2.1.6 Energy of a continuous distribution	2-2
		2.1.7 Energy per unit volume	2-2
	2.2	Problems and solutions	2-3
		Bibliography	2-35

3 Electric potential 3-1

3.1	Theory	3-1
	3.1.1 Laplace's equation	3-1
	3.1.2 Solving Laplace's equation	3-1
	3.1.3 General solutions	3-4
	3.1.4 Method of images	3-5
	3.1.5 Potential due to a dipole	3-6
	3.1.6 Multiple expansion	3-6
	3.1.7 Monopole moment	3-6
3.2	Problems and solutions	3-6
	Bibliography	3-29

4 Magnetostatics 4-1

4.1	Theory	4-1
	4.1.1 Magnetic force	4-1
	4.1.2 Force on a current carrying wire	4-1
	4.1.3 Volume current density	4-1
	4.1.4 Continuity equation	4-2
	4.1.5 Biot–Savart law	4-2
	4.1.6 Divergence of $\vec{B}$	4-2
	4.1.7 Ampère's law	4-2
	4.1.8 Vector potential	4-2
	4.1.9 Magnetic dipole moment	4-3
	4.1.10 Magnetic field due to dipole moment	4-3
4.2	Problems and solutions	4-3
	Bibliography	4-26

5 Electric fields in matter 5-1

5.1	Theory	5-1
	5.1.1 Induced dipole moment of an atom in an electric field	5-1
	5.1.2 Torque on a dipole due to an electric field	5-1
	5.1.3 Force on a dipole	5-1

5.1.4	Energy of a dipole in an electric field	5-2
5.1.5	Surface bound charge due to polarization $\vec{P}$	5-2
5.1.6	Volume bound charge due to polarization $\vec{P}$	5-2
5.1.7	Potential due to polarization $\vec{P}$	5-2
5.1.8	Electric displacement	5-2
5.1.9	Gauss's law for electric displacement	5-2
5.1.10	Linear dielectrics	5-2
5.1.11	Energy in a dielectric system	5-3
5.2 Problems and solutions		5-3
Bibliography		5-26

6 Magnetic fields in matter 6-1

6.1 Theory		6-1
6.1.1	Torque on a magnetic dipole moment	6-1
6.1.2	Force on a magnetic dipole	6-1
6.1.3	H-field	6-1
6.1.4	Linear media	6-2
6.1.5	Surface bound current due to magnetization $\vec{M}$	6-2
6.1.6	Volume bound current due to magnetization $\vec{M}$	6-2
6.2 Problems and solutions		6-2
Bibliography		6-14

编辑手记　　　　　　　　　　　　　　　　　　　　　　　　E-1

Preface

We wrote this book of problems and solutions having in mind the undergraduate student—sophomore, junior, or senior—who may want to work on more problems and receive immediate feedback while studying. The authors strongly recommend the textbook by David J Griffiths, *Introduction to Electrodynamics*, as a first source manual, since it is recognized as one of the best books on electrodynamics at the undergraduate level. We consider this book of problems and solutions a companion volume for the student who would like to work on more electrostatic problems by herself/himself in order to deepen their understanding and problems solving skills. We add brief theoretical notes and formulae; for a complete theoretical approach we suggest Griffiths' book. Every chapter is organized as follows: brief theoretical notes followed by the problem text with the solution. Each chapter ends with a brief bibliography.

We plan to write a second volume on electrodynamics, which will start with Maxwell's equations and the conservation laws, and then discuss electromagnetic (EM) waves, potentials and fields, radiation, and relativistic electrodynamics.

We follow here the notation of Griffiths, and use $\vec{r}$ for the vector from a source point $\vec{r}'$ to the field point $\vec{r}$. Please note that $\hat{r} = \dfrac{\vec{r}}{r} = \dfrac{\vec{r} - \vec{r}'}{|\vec{r} - \vec{r}'|}$ and, as you see, this notation already greatly simplifies complex equations, but you need to be careful with your notation, in particular if you only use cursive or typed letters. Also, we use the same notation s for the distance to the z-axis in cylindrical coordinates as is used in Griffiths' book.

The chosen units are SI units—the international system. The reader should be aware that other books may employ either the Gaussian system (CGS) or the Heaviside–Lorentz (HL) system. The Coulomb force in each of the systems is as follows,

SI system:
$$\vec{F} = \frac{1}{4\pi\varepsilon_0} \frac{q_1 q_2}{r^2} \hat{r}$$

CGS:
$$\vec{F} = \frac{q_1 q_2}{r^2} \hat{r}$$

HL:
$$\vec{F} = \frac{1}{4\pi} \frac{q_1 q_2}{r^2} \hat{r}$$

Some of the problems are typical practice problems with the pedagogical role of improving understanding and problem solving skills. Several of the problems presented here appear in a variety of undergraduate textbooks on EM as they are classic examples; however, we felt it would be incomplete to omit these problems as they are fundamental to the study of EM. We also present problems that are more general in nature, which may be a bit more challenging. We tried to maintain a balance between the two types of problems, and we hope that the readers will enjoy this variation and have as much thrill and excitement as we had while creating and solving these problems.

Acknowledgements

We want to thank to Dr Ilie's students, Nicholas Jira, Vincent DeBiase, Ian Evans, and Andres Inga, who contributed to the editing (typing) of this book. We are particularly grateful to our illustrator, Julia D'Rozario, for making all of the figures. We thank Dr Ildar Sabirianov for providing useful suggestions. We thank the administration at SUNY Oswego and the office of Research and Individualized Student Experiences for overall support. We are grateful to Dr Peter Dowben, from the University of Nebraska at Lincoln, who thought that such a project has a niche. A thought of appreciation to Dr Charles Ebner, from the Ohio State University for his perfect Electrodynamics course. Also many thanks to our editors, Joel Claypool, Publisher at Morgan & Claypool Publishers, Jeanine Burke, Consulting Editor at the IOP Concise Physics e-book program, and Jacky Mucklow, Production Team Manager at the Institute of Physics. Lastly, we thank to our families and friends for their sense of humor, encouragement, and for keeping us sane and happy.

About the authors

Carolina C Ilie

Carolina C Ilie is an Associate Professor with tenure at the State University of New York at Oswego. She taught Electromagnetic Theory for almost ten years and designed various problems for her students' exams, group work, and quizzes. Dr Ilie obtained her PhD in Physics and Astronomy at the University of Nebraska at Lincoln, an MSc in Physics at Ohio State University and another MSc in Physics at the University of Bucharest, Romania. She received the President's Award for Teaching Excellence in 2016 and the Provost Award for Mentoring in Scholarly and Creative Activity in 2013. She lives in Central New York with her spouse, also a physicist, and their two sons.

Photograph courtesy of James Russell/SUNY Oswego Office of Communications and Marketing.

Zachariah S Schrecengost

Zachariah S Schrecengost is a State University of New York alumnus. He graduated summa cum laude with a BS degree having completed majors in Physics, Software Engineering, and Applied Mathematics. He took the Advanced Electromagnetic Theory course with Dr Ilie and was thrilled to be involved in creating this book. He brings to the project both the fresh perspective of the student taking electrodynamics, as well as the enthusiasm and talent of an alumnus who is an electrodynamics and upper level mathematics aficionado. Mr Schrecengost works as a software engineer in Syracuse and is preparing to begin his graduate school studies in physics.

Julia R D'Rozario

Julia R D'Rozario (*illustrator*) graduated from the State University of New York at Oswego in December 2016 where she completed a BS in Physics and a BA in Cinema and Screen Studies, and completed a minor in Astronomy by May 2016. She completed the Advanced Electromagnetic Theory course with Dr Ilie and has much experience of the arts through her career in film. Ms D'Rozario contributes her knowledge of electrodynamics and her talent in drawing using Inkscape software. Her future aim is to attend graduate school and continue to combine her passions for physics and cinema.

IOP Concise Physics

Electromagnetism
Problems and solutions

Carolina C Ilie and Zachariah S Schrecengost

Chapter 1

Mathematical techniques

There are a variety of mathematical techniques required to solve problems in electromagnetism. The aim of this chapter is to provide problems that will build confidence in these techniques. Concepts from vector calculus and curvilinear coordinate systems are the primary focus.

1.1 Theory

1.1.1 Dot and cross product

Given vectors $\vec{A} = A_x\hat{x} + A_y\hat{y} + A_z\hat{z}$ and $\vec{B} = B_x\hat{x} + B_y\hat{y} + B_z\hat{z}$

$$\vec{A} \cdot \vec{B} = A_xB_x + A_yB_y + A_zB_z = AB\cos\theta$$

$$\vec{A} \times \vec{B} = \begin{vmatrix} \hat{x} & \hat{y} & \hat{z} \\ A_x & A_y & A_z \\ B_x & B_y & B_z \end{vmatrix} \text{ with } |\vec{A} \times \vec{B}| = AB\sin\theta$$

where $A = |\vec{A}| = \sqrt{A_x^2 + A_y^2 + A_z^2}$, $B = |\vec{B}| = \sqrt{B_x^2 + B_y^2 + B_z^2}$, and θ is the angle between $\vec{A}$ and $\vec{B}$.

1.1.2 Separation vector

This notation is outlined by David J Griffiths in his book *Introduction to Electrodynamics* (1999, 2013). Given a source point $\vec{r}'$ and field point $\vec{r}$, the separation vector points from $\vec{r}'$ to $\vec{r}$ and is given by

$$\vec{\imath} = \vec{r} - \vec{r}' = (x - x')\hat{x} + (y - y')\hat{y} + (z - z')\hat{z}$$

and the unit vector pointing from $\vec{r}'$ to $\vec{r}$ is

$$\hat{r} = \frac{\vec{r}}{r} = \frac{\vec{r}-\vec{r}'}{|\vec{r}-\vec{r}'|} = \frac{(x-x')\hat{x} + (y-y')\hat{y} + (z-z')\hat{z}}{\sqrt{(x-x')^2 + (y-y')^2 + (z-z')^2}}.$$

As explained by Griffiths, this notation greatly simplifies later equations.

1.1.3 Transformation matrix

Given vector $\vec{A} = A_x\hat{x} + A_y\hat{y} + A_z\hat{z}$ in coordinate system K, the components of $\vec{A}$ in coordinate system K' are determined by rotational matrix R given by

$$R = \begin{pmatrix} R_{xx} & R_{xy} & R_{xz} \\ R_{yx} & R_{yy} & R_{yz} \\ R_{zx} & R_{zy} & R_{zz} \end{pmatrix}$$

with

$$\begin{pmatrix} A'_x \\ A'_y \\ A'_z \end{pmatrix} = R \begin{pmatrix} A_x \\ A_y \\ A_z \end{pmatrix}.$$

1.1.4 Gradient

Given a scalar function T, the gradients for various coordinate systems are given below.

Cartesian

$$\nabla T = \frac{\partial T}{\partial x}\hat{x} + \frac{\partial T}{\partial y}\hat{y} + \frac{\partial T}{\partial z}\hat{z}$$

Cylindrical

$$\nabla T = \frac{\partial T}{\partial s}\hat{s} + \frac{1}{s}\frac{\partial T}{\partial \phi}\hat{\phi} + \frac{\partial T}{\partial z}\hat{z}$$

Spherical

$$\nabla T = \frac{\partial T}{\partial r}\hat{r} + \frac{1}{r}\frac{\partial T}{\partial \theta}\hat{\theta} + \frac{1}{r\sin\theta}\frac{\partial T}{\partial \phi}\hat{\phi}$$

1.1.5 Divergence

Given vector function $\vec{v}$, the divergences for various coordinate systems are given below.

Cartesian

$$\nabla \cdot \vec{v} = \frac{\partial v_x}{\partial x} + \frac{\partial v_y}{\partial y} + \frac{\partial v_z}{\partial z}$$

Cylindrical

$$\nabla \cdot \vec{v} = \frac{1}{s}\frac{\partial}{\partial s}(sv_s) + \frac{1}{s}\frac{\partial v_\phi}{\partial \phi} + \frac{\partial v_z}{\partial z}$$

Spherical

$$\nabla \cdot \vec{v} = \frac{1}{r^2}\frac{\partial}{\partial r}(r^2 v_r) + \frac{1}{r\sin\theta}\frac{\partial}{\partial \theta}(\sin\theta\, v_\theta) + \frac{1}{r\sin\theta}\frac{\partial v_\phi}{\partial \phi}$$

1.1.6 Curl

Given vector function $\vec{v}$, the curls for various coordinate systems are given below.

Cartesian

$$\nabla \times \vec{v} = \left(\frac{\partial v_z}{\partial y} - \frac{\partial v_y}{\partial z}\right)\hat{x} + \left(\frac{\partial v_x}{\partial z} - \frac{\partial v_z}{\partial x}\right)\hat{y} + \left(\frac{\partial v_y}{\partial x} - \frac{\partial v_x}{\partial y}\right)\hat{z}$$

Cylindrical

$$\nabla \times \vec{v} = \left(\frac{1}{s}\frac{\partial v_z}{\partial \phi} - \frac{\partial v_\phi}{\partial z}\right)\hat{s} + \left(\frac{\partial v_s}{\partial z} - \frac{\partial v_z}{\partial s}\right)\hat{\phi} + \frac{1}{s}\left[\frac{\partial}{\partial s}(sv_\phi) - \frac{\partial v_s}{\partial \phi}\right]\hat{z}$$

Spherical

$$\nabla \times \vec{v} = \frac{1}{r\sin\theta}\left[\frac{\partial}{\partial \theta}(\sin\theta\, v_\phi) - \frac{\partial v_\theta}{\partial \phi}\right]\hat{r} + \frac{1}{r}\left[\frac{1}{\sin\theta}\frac{\partial v_r}{\partial \phi} - \frac{\partial}{\partial r}(rv_\phi)\right]\hat{\theta}$$
$$+ \frac{1}{r}\left[\frac{\partial}{\partial r}(rv_\theta) - \frac{\partial v_r}{\partial \theta}\right]\hat{\phi}$$

1.1.7 Laplacian

Given a scalar function T, the Laplacians for various coordinate systems are given below.

Cartesian

$$\nabla^2 T = \frac{\partial^2 T}{\partial x^2} + \frac{\partial^2 T}{\partial y^2} + \frac{\partial^2 T}{\partial z^2}$$

Cylindrical

$$\nabla^2 T = \frac{1}{s}\frac{\partial}{\partial s}\left(s\frac{\partial T}{\partial s}\right) + \frac{1}{s^2}\frac{\partial^2 T}{\partial \phi^2} + \frac{\partial^2 T}{\partial z^2}$$

Spherical

$$\nabla^2 T = \frac{1}{r^2}\frac{\partial}{\partial r}\left(r^2\frac{\partial T}{\partial r}\right) + \frac{1}{r^2 \sin\theta}\frac{\partial}{\partial \theta}\left(\sin\theta\frac{\partial T}{\partial \theta}\right) + \frac{1}{r^2 \sin^2\theta}\frac{\partial^2 T}{\partial \phi^2}$$

1.1.8 Line integral

Given vector function $\vec{v}$ and path $\mathcal{P}$, a line integral is given by

$$\int_{\vec{a}\,\mathcal{P}}^{\vec{b}} \vec{v} \cdot \mathrm{d}\vec{\ell},$$

where $\vec{a}$ and $\vec{b}$ are the end points, and $\mathrm{d}\vec{\ell}$ is the infinitesimal displacement vector along $\mathcal{P}$. In Cartesian coordinates $\mathrm{d}\vec{\ell} = \mathrm{d}x\,\hat{x} + \mathrm{d}y\,\hat{y} + \mathrm{d}z\,\hat{z}$.

1.1.9 Surface integral

Given vector function $\vec{v}$ and surface $\mathcal{S}$, a surface integral is given by

$$\int_{\mathcal{S}} \vec{v} \cdot \mathrm{d}\vec{a},$$

where $\mathrm{d}\vec{a}$ is the infinitesimal area vector that has direction normal to the surface. Note that $\mathrm{d}\vec{a}$ always depends on the surface involved.

1.1.10 Volume integral

Given scalar function T and volume $\mathcal{V}$, a volume integral is given by

$$\int_{\mathcal{V}} T\,\mathrm{d}\tau,$$

where $\mathrm{d}\tau$ is the infinitesimal volume element. In Cartesian coordinates $\mathrm{d}\tau = \mathrm{d}x\,\mathrm{d}y\,\mathrm{d}z$.

1.1.11 Fundamental theorem for gradients

$$\int_{\vec{a}\,\mathcal{P}}^{\vec{b}} (\nabla T) \cdot \mathrm{d}\vec{\ell} = T(\vec{b}) - T(\vec{a})$$

1.1.12 Fundamental theorem for divergences (Gauss's theorem, Green's theorem, divergence theorem)

$$\int_{\mathcal{V}} (\nabla \cdot \vec{v})\,\mathrm{d}\tau = \oint_{\mathcal{S}} \vec{v} \cdot \mathrm{d}\vec{a}$$

1.1.13 Fundamental theorem for curls (Stoke's theorem, curl theorem)

$$\int_{\mathcal{S}} (\nabla \times \vec{v}) \cdot \mathrm{d}\vec{a} = \oint_{\mathcal{P}} \vec{v} \cdot \mathrm{d}\vec{\ell}$$

1.1.14 Cylindrical polar coordinates

Here our infinitesimal quantities are

$$\mathrm{d}\vec{\ell} = \mathrm{d}s\,\hat{s} + s\,\mathrm{d}\phi\,\hat{\phi} + \mathrm{d}z\,\hat{z}$$

and

$$\mathrm{d}\tau = s\,\mathrm{d}s\,\mathrm{d}\phi\,\mathrm{d}z.$$

1.1.15 Spherical polar coordinates

Here our infinitesimal quantities are

$$d\vec{\ell} = dr\,\hat{r} + r\,d\theta\,\hat{\theta} + r\sin\theta\,d\phi\,\hat{\phi}$$

and

$$d\tau = r^2 \sin\theta\,dr\,d\theta\,d\phi.$$

1.1.16 One-dimensional Dirac delta function

The one-dimensional Dirac delta function is given by

$$\delta(x-a) = \begin{cases} 0 & x \neq a \\ \infty & x = a \end{cases}$$

and has the following properties

$$\int_{-\infty}^{\infty} \delta(x-a)\,dx = 1$$

$$\int_{-\infty}^{\infty} f(x)\delta(x-a)\,dx = f(a)$$

$$\delta(kx) = \frac{1}{|k|}\delta(x).$$

1.1.17 Theory of vector fields

If the curl of a vector field $\vec{F}$ vanishes everywhere, then $\vec{F}$ can be written as the gradient of a scalar potential V:

$$\nabla \times \vec{F} \leftrightarrow \vec{F} = -\nabla V.$$

If the divergence of a vector vanishes everywhere, then $\vec{F}$ can be expressed as the curl of a vector potential $\vec{A}$:

$$\nabla \cdot \vec{F} = 0 \leftrightarrow \vec{F} = \nabla \times \vec{A}.$$

1.2 Problems and solutions

Problem 1.1. Given vectors $\vec{A} = 3\hat{x} + 9\hat{y} + 5\hat{z}$ and $\vec{B} = \hat{x} - 7\hat{y} + 4\hat{z}$, calculate $\vec{A} \cdot \vec{B}$ and $\vec{A} \times \vec{B}$ using vector components and find the angle between $\vec{A}$ and $\vec{B}$ using both products.

Solution

$$\vec{A} \cdot \vec{B} = (3\hat{x} + 9\hat{y} + 5\hat{z}) \cdot (\hat{x} - 7\hat{y} + 4\hat{z})$$
$$= (3)(1) + (9)(-7) + (5)(4) = 3 - 63 + 20$$
$$\vec{A} \cdot \vec{B} = -40$$

$$\vec{A} \times \vec{B} = \begin{vmatrix} \hat{x} & \hat{y} & \hat{z} \\ 3 & 9 & 5 \\ 1 & -7 & 4 \end{vmatrix}$$
$$= [(9)(4) - (-7)(5)]\hat{x} + [(1)(5) - (3)(4)]\hat{y} + [(3)(-7) - (1)(9)]\hat{z}$$
$$\hat{A} \times \hat{B} = 71\hat{x} - 7\hat{y} - 30\hat{z}$$

To find the angle θ between $\vec{A}$ and $\hat{B}$ we must first calculate A and B:

$$A = \sqrt{3^2 + 9^2 + 5^2} = \sqrt{115}$$
$$B = \sqrt{1^2 + (-7)^2 + 4^2} = \sqrt{66}.$$

Using the dot product, we have

$$\vec{A} \cdot \vec{B} = AB \cos\theta \rightarrow \theta = \cos^{-1}\left(\frac{-40}{\sqrt{115}\sqrt{66}}\right)$$
$$\theta = 117.3°.$$

Using the cross product, we have

$$\left|\vec{A} \times \vec{B}\right| = AB \sin\theta \rightarrow \sqrt{71^2 + (-7)^2 + (-30)^2} = \sqrt{115}\sqrt{66}\sin\theta$$
$$\theta = 62.7°.$$

Note, however, that we can see that the angle between $\vec{A}$ and $\vec{B}$ is greater than 90°. For any argument γ, $-90° \leqslant \sin^{-1}(\gamma) \leqslant 90°$. Since the angle between $\vec{A}$ and $\vec{B}$ is greater than 90°, we must adjust for this by subtracting our angle from 180°. Therefore, $\theta = 180° - 62.7° = 117.3°$ as expected.

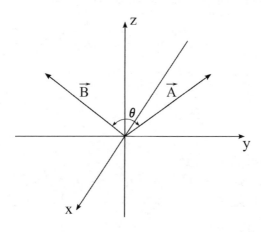

Problem 1.2. The scalar triple product states $\vec{A} \cdot (\vec{B} \times \vec{C}) = \vec{B} \cdot (\vec{C} \times \vec{A})$. Prove this by expressing each side in terms of its components.

Solution Starting with the left-hand side, the cross product is

$$\vec{B} \times \vec{C} = \begin{vmatrix} \hat{x} & \hat{y} & \hat{z} \\ B_x & B_y & B_z \\ C_x & C_y & C_z \end{vmatrix}$$

$$= (B_y C_z - B_z C_y)\hat{x} + (B_z C_x - B_x C_z)\hat{y} + (B_x C_y - B_y C_x)\hat{z}.$$

Now, dotting $\vec{A}$ with $(\vec{B} \times \vec{C})$

$$\vec{A} \cdot (\vec{B} \times \vec{C}) = A_x(B_y C_z - B_z C_y) + A_y(B_z C_x - B_x C_z) + A_z(B_x C_y - B_y C_x)$$

$$= A_x B_y C_z - A_x B_z C_y + A_y B_z C_x - A_y B_x C_z + A_z B_x C_y - A_z B_y C_x$$

$$= B_x(C_y A_z - C_z A_y) + B_y(C_z A_x - C_x A_z) + B_z(C_x A_y - C_y A_x)$$

$$\vec{A} \cdot (\vec{B} \times \vec{C}) = \vec{B} \cdot \left[(C_y A_z - C_z A_y)\hat{x} + (C_z A_x - C_x A_z)\hat{y} + (C_x A_y - C_y A_x)\hat{z} \right].$$

Note the term in brackets is precisely $\vec{C} \times \vec{A}$, therefore

$$\vec{A} \cdot (\vec{B} \times \vec{C}) = \vec{B} \cdot (\vec{C} \times \vec{A})$$

as desired. This procedure can easily be applied again to prove the final part of the triple product,

$$\vec{A} \cdot (\vec{B} \times \vec{C}) = \vec{B} \cdot (\vec{C} \times \vec{A}) = \vec{C} \cdot (\vec{A} \times \vec{B}).$$

Problem 1.3. Given source vector $\vec{r}' = r \cos\theta \, \hat{x} + r \sin\theta \, \hat{y}$ and field vector $\vec{r} = z\hat{z}$, find the separation vector $\vec{\mathfrak{r}}$ and the unit vector $\hat{\mathfrak{r}}$.

Solution We have

$$\vec{\mathfrak{r}} = \vec{r} - \vec{r}' = z\hat{z} - \left(r \cos\theta \, \hat{x} + r \sin\theta \, \hat{y} \right)$$

$$\vec{\mathfrak{r}} = -r \cos\theta \, \hat{x} - r \sin\theta \, \hat{y} + z\hat{z}.$$

To determine the unit vector $\hat{\mathfrak{r}}$, we must first find the magnitude of $\vec{\mathfrak{r}}$,

$$\mathfrak{r} = \sqrt{(-r\cos\theta)^2 + (-r\sin\theta)^2 + z^2} = \sqrt{r^2(\cos^2\theta + \sin^2\theta) + z^2} = \sqrt{r^2 + z^2}.$$

So
$$\hat{r} = \frac{\vec{r}}{r} = \frac{-r\cos\theta\,\hat{x} - r\sin\theta\,\hat{y} + z\hat{z}}{\sqrt{r^2 + z^2}}.$$

Problem 1.4. Given $\vec{A}$ in coordinate system K, find the rotational matrix to give the components in system K'.

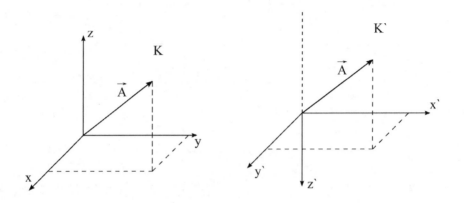

Solution From the figures, we have

$$A'_x = A_y, \qquad A'_y = A_x, \qquad A'_z = -A_z.$$

We want to find the rotational matrix R that satisfies

$$\begin{pmatrix} A'_x \\ A'_y \\ A'_z \end{pmatrix} = R \begin{pmatrix} A_x \\ A_y \\ A_z \end{pmatrix}.$$

From our equations above

$$\begin{pmatrix} A'_x \\ A'_y \\ A'_z \end{pmatrix} = \begin{pmatrix} 0 & 1 & 0 \\ 1 & 0 & 0 \\ 0 & 0 & -1 \end{pmatrix} \begin{pmatrix} A_x \\ A_y \\ A_z \end{pmatrix}.$$

Therefore,
$$R = \begin{pmatrix} 0 & 1 & 0 \\ 1 & 0 & 0 \\ 0 & 0 & -1 \end{pmatrix}.$$

Problem 1.5. Find the gradient of the following functions:
a) $T = x^4 + y^2 + z^3$
b) $T = x^2 \ln y \, z^3$
c) $T = x^2 y + z^3$

Solutions
a) $T = x^4 + y^2 + z^3$

$$\nabla T = \frac{\partial T}{\partial x}\hat{x} + \frac{\partial T}{\partial y}\hat{y} + \frac{\partial T}{\partial z}\hat{z} = 4x^3\hat{x} + 2y\hat{y} + 3z^2\hat{z}$$

b) $T = x^2 \ln y \, z^3$

$$\nabla T = \frac{\partial T}{\partial x}\hat{x} + \frac{\partial T}{\partial y}\hat{y} + \frac{\partial T}{\partial z}\hat{z} = 2xz^3 \ln y \, \hat{x} + \frac{x^2 z^3}{y}\hat{y} + 3x^2 z^2 \ln y \, \hat{z}$$

c) $T = x^2 y + z^3$

$$\nabla T = \frac{\partial T}{\partial x}\hat{x} + \frac{\partial T}{\partial y}\hat{y} + \frac{\partial T}{\partial z}\hat{z} = 2xy\hat{x} + x^2\hat{y} + 3z^2\hat{z}$$

Problem 1.6. Find the divergence of the following functions:
a) $\vec{v} = xy\hat{x} - 2y^2 z\hat{y} + z^3\hat{z}$
b) $\vec{v} = (x + y)\hat{x} + (y + z)\hat{y} + (z + x)\hat{z}$

Solutions
a) $\vec{v} = xy\hat{x} - 2y^2 z\hat{y} + z^3\hat{z}$

$$\nabla \cdot \vec{v} = \frac{\partial v_x}{\partial x} + \frac{\partial v_y}{\partial y} + \frac{\partial v_z}{\partial z} = y - 4yz + 3z^2$$

b) $\vec{v} = (x + y)\hat{x} + (y + z)\hat{y} + (z + x)\hat{z}$

$$\nabla \cdot \vec{v} = \frac{\partial v_x}{\partial x} + \frac{\partial v_y}{\partial y} + \frac{\partial v_z}{\partial z} = 1 + 1 + 1 = 3$$

Problem 1.7. Find the curl of the following functions:
a) $\vec{v} = xy\hat{x} - 2y^2 z\hat{y} + z^3\hat{z}$
b) $\vec{v} = (x + y)\hat{x} + (y + z)\hat{y} + (z + x)\hat{z}$
c) $\vec{v} = \sin x \, \hat{x} + \cos y \, \hat{y}$

Solutions
a) $\vec{v} = xy\hat{x} - 2y^2z\hat{y} + z^3\hat{z}$

$$\nabla \times \vec{v} = \left(\frac{\partial v_z}{\partial y} - \frac{\partial v_y}{\partial z}\right)\hat{x} + \left(\frac{\partial v_x}{\partial z} - \frac{\partial v_z}{\partial x}\right)\hat{y} + \left(\frac{\partial v_y}{\partial x} - \frac{\partial v_x}{\partial y}\right)\hat{z}$$

$$= \left[\frac{\partial(z^3)}{\partial y} - \frac{\partial(-2y^2z)}{\partial z}\right]\hat{x} + \left[\frac{\partial(xy)}{\partial z} - \frac{\partial(z^3)}{\partial x}\right]\hat{y}$$

$$+ \left[\frac{\partial(-2y^2z)}{\partial x} - \frac{\partial(xy)}{\partial y}\right]\hat{z}$$

$$= (0 + 2y^2)\hat{x} + (0 - 0)\hat{y} + (0 - x)\hat{z}$$

$$\nabla \times \vec{v} = 2y^2\hat{x} - x\hat{z}$$

b) $\vec{v} = (x + y)\hat{x} + (y + z)\hat{y} + (z + x)\hat{z}$

$$\nabla \times \vec{v} = \left(\frac{\partial v_z}{\partial y} - \frac{\partial v_y}{\partial z}\right)\hat{x} + \left(\frac{\partial v_x}{\partial z} - \frac{\partial v_z}{\partial x}\right)\hat{y} + \left(\frac{\partial v_y}{\partial x} - \frac{\partial v_x}{\partial y}\right)\hat{z}$$

$$= \left[\frac{\partial(z+x)}{\partial y} - \frac{\partial(y+z)}{\partial z}\right]\hat{x} + \left[\frac{\partial(x+y)}{\partial z} - \frac{\partial(z+x)}{\partial x}\right]\hat{y}$$

$$+ \left[\frac{\partial(y+z)}{\partial x} - \frac{\partial(x+y)}{\partial y}\right]\hat{z}$$

$$\nabla \times \vec{v} = -\hat{x} - \hat{y} - \hat{z}$$

c) $\vec{v} = \sin x\,\hat{x} + \cos y\,\hat{y}$

$$\nabla \times \vec{v} = \left(\frac{\partial v_z}{\partial y} - \frac{\partial v_y}{\partial z}\right)\hat{x} + \left(\frac{\partial v_x}{\partial z} - \frac{\partial v_z}{\partial x}\right)\hat{y} + \left(\frac{\partial v_y}{\partial x} - \frac{\partial v_x}{\partial y}\right)\hat{z}$$

$$= \left[\frac{\partial(0)}{\partial y} - \frac{\partial(\cos y)}{\partial z}\right]\hat{x} + \left[\frac{\partial(\sin x)}{\partial z} - \frac{\partial(0)}{\partial x}\right]\hat{y}$$

$$+ \left[\frac{\partial(\cos y)}{\partial x} - \frac{\partial(\sin x)}{\partial y}\right]\hat{z} = 0$$

Problem 1.8. Prove $\nabla \times (\nabla T) = 0$.

Solution

$$\nabla \times (\nabla T) = \begin{vmatrix} \hat{x} & \hat{y} & \hat{z} \\ \dfrac{\partial}{\partial x} & \dfrac{\partial}{\partial y} & \dfrac{\partial}{\partial z} \\ \dfrac{\partial T}{\partial x} & \dfrac{\partial T}{\partial y} & \dfrac{\partial T}{\partial z} \end{vmatrix}$$

$$= \left[\frac{\partial}{\partial y}\left(\frac{\partial T}{\partial z}\right) - \frac{\partial}{\partial z}\left(\frac{\partial T}{\partial y}\right) \right]\hat{x} + \left[\frac{\partial}{\partial z}\left(\frac{\partial T}{\partial x}\right) - \frac{\partial}{\partial x}\left(\frac{\partial T}{\partial z}\right) \right]\hat{y}$$
$$+ \left[\frac{\partial}{\partial x}\left(\frac{\partial T}{\partial y}\right) - \frac{\partial}{\partial y}\left(\frac{\partial T}{\partial x}\right) \right]\hat{z}$$

$$\nabla \times (\nabla T) = 0.$$

Problem 1.9. Find the Laplacian of the following functions:
a) $T = x + y^2 + xz + 3$
b) $T = e^x + \sin y \cos(2z)$
c) $T = \sin x \cos y$
d) $\vec{v} = xy\hat{x} + z^2\hat{y} - 2\hat{z}$

Solutions
a) $T = x + y^2 + xz + 3$

$$\nabla^2 T = \frac{\partial^2 T}{\partial x^2} + \frac{\partial^2 T}{\partial y^2} + \frac{\partial^2 T}{\partial z^2} = 0 + 2 + 0 = 2$$

b) $T = e^x + \sin y \cos(2z)$

$$\nabla^2 T = \frac{\partial^2 T}{\partial x^2} + \frac{\partial^2 T}{\partial y^2} + \frac{\partial^2 T}{\partial z^2} = e^x - \sin y \cos(2z) - 4 \sin y \cos(2z)$$

$$= e^x - 5 \sin y \cos(2z)$$

c) $T = \sin x \cos y$

$$\nabla^2 T = \frac{\partial^2 T}{\partial x^2} + \frac{\partial^2 T}{\partial y^2} + \frac{\partial^2 T}{\partial z^2} = -\sin x \cos y - \sin x \cos y = -2 \sin x \cos y$$

d) $\vec{v} = xy\hat{x} + z^2\hat{y} - 2\hat{z}$

$$\nabla^2 \vec{v} = \left(\frac{\partial^2 v_x}{\partial x^2} + \frac{\partial^2 v_x}{\partial y^2} + \frac{\partial^2 v_x}{\partial z^2}\right)\hat{x} + \left(\frac{\partial^2 v_y}{\partial x^2} + \frac{\partial^2 v_y}{\partial y^2} + \frac{\partial^2 v_y}{\partial z^2}\right)\hat{y}$$
$$+ \left(\frac{\partial^2 v_z}{\partial x^2} + \frac{\partial^2 v_z}{\partial y^2} + \frac{\partial^2 v_z}{\partial z^2}\right)\hat{z}$$

$$\nabla^2 \vec{v} = (0 + 0 + 0)\hat{x} + (0 + 0 + 2)\hat{y} + (0 + 0 + 0)\hat{z} = 2\hat{y}$$

Problem 1.10. Test the divergence theorem with $\vec{v} = 2xy\hat{x} + y^2z^3\hat{y} + (x^2z - 2y)\hat{z}$ and the volume below.

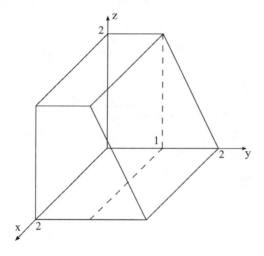

Solution The divergence theorem states

$$\int_V \nabla \cdot \vec{v} \, d\tau = \oint_S \vec{v} \cdot d\vec{a}.$$

Starting with the left-hand side, we have the divergence

$$\nabla \cdot \vec{v} = 2y + 2yz^3 + x^2 = 2y\left(z^3 + 1\right) + x^2.$$

We must split the volume into two pieces, (a) $0 \leqslant y \leqslant 1$ and (b) $1 \leqslant y \leqslant 2$.

(a)

$$\int_V \nabla \cdot \vec{v} \, d\tau = \int_0^2 \int_0^2 \int_0^1 \left[2y\left(z^3 + 1\right) + x^2\right] dy \, dx \, dz = \frac{52}{3}$$

(b)

$$\int \nabla \cdot \vec{v} \, d\tau = \int_0^2 \int_1^2 \int_0^{4-2y} \left[2y(z^3 + 1) + x^2 \right] dy \, dx \, dz = \frac{176}{15}$$

So,

$$\int_v \nabla \cdot \vec{v} \, d\tau = \frac{52}{3} + \frac{176}{15} = \frac{436}{15}.$$

Now we solve $\oint_S \vec{v} \cdot d\vec{a}$, which must be evaluated over the six sides.

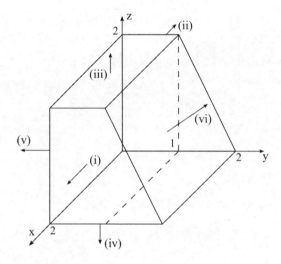

(i) We must split this region into two sections (a) and (b), and $d\vec{a} = dy \, dz \, \hat{z}$ with $x = 2$.

In (a), $0 \leqslant y \leqslant 1$,

$$\int \vec{v} \cdot d\vec{a} = \int_0^2 \int_0^1 2(2)y \, dy \, dz = 4.$$

In (b), $1 \leqslant y \leqslant 2$ and $0 \leqslant z \leqslant 4 - 2y$

$$\int \vec{v} \cdot d\vec{a} = \int_1^2 \int_0^{4-2y} 2(2)y \, dz \, dy = \frac{16}{3}.$$

(ii) Here, $d\vec{a} = -dy\, dz\, \hat{x}$ and $x = 0$, so $\vec{v} \cdot d\vec{a} = 2(0)y(-dy\, dx)\hat{x} = 0$.
(iii) Here, $d\vec{a} = dx\, dy\, \hat{z}$ and $z = 2$

$$\int \vec{v} \cdot d\vec{a} = \int_0^2 \int_0^1 \left[x^2(2) - 2y \right] dy\, dx = \frac{10}{3}.$$

(iv) Here, $d\vec{a} = -dx\, dy\, \hat{z}$ and $z = 0$

$$\int \vec{v} \cdot d\vec{a} = \int_0^2 \int_0^2 \left[x^2(0) - 2y \right](-dx\, dy) = 8.$$

(v) Here $d\vec{a} = dx\, dz\, \hat{y}$ and $y = 0$, so $\vec{v} \cdot d\vec{a} = 0^2 z^3(-dx\, dz) = 0$.
(vi) Here, we have $d\vec{a} = dx\, dz'\, \hat{n}$ where $\hat{n} = \frac{\vec{n}}{n}$. We can find $\vec{n}$ by crossing vectors $\vec{A} = -\hat{y} + 2\hat{z}$ and $\vec{B} = 2\hat{x}$ (the edges of the volume):

$$\vec{n} = \vec{A} \times \vec{B} = \begin{vmatrix} \hat{x} & \hat{y} & \hat{z} \\ 0 & -1 & 2 \\ 2 & 0 & 0 \end{vmatrix} = 4\hat{y} + 2\hat{z}.$$

So

$$n = \sqrt{4^2 + 2^2} = 2\sqrt{5}$$

and

$$\hat{n} = \frac{2\sqrt{5}}{5}\hat{y} + \frac{\sqrt{5}}{5}\hat{z}.$$

We can also obtain dz' by considering

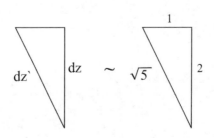

so $dz' = \frac{\sqrt{5}}{2}dz$. Now

$$d\vec{a} = \frac{\sqrt{5}}{2}dx\,dz\left(\frac{2\sqrt{5}}{5}\hat{y} + \frac{\sqrt{5}}{5}\hat{z}\right) = \left(\hat{y} + \frac{1}{2}\hat{z}\right)dx\,dz$$

and

$$z = 4 - 2y \rightarrow y = 2 - \frac{z}{2}.$$

So

$$\int \vec{v} \cdot d\vec{a} = \int_0^2\int_0^2 \left[y^2z^3 + \frac{1}{2}(x^2z - 2y)\right]dx\,dz$$

$$= \int_0^2\int_0^2 \left\{\left(2 - \frac{z}{2}\right)^2 z^3 + \frac{1}{2}\left[x^2z - 2\left(2 - \frac{z}{2}\right)\right]\right\}dx\,dz = \frac{42}{5}.$$

Therefore

$$\oint_S \vec{v} \cdot d\vec{a} = 4 + \frac{16}{3} + \frac{10}{3} + 8 + \frac{42}{5} = \frac{436}{15}$$

as expected.

Problem 1.11. Test the curl theorem with $\vec{v} = 5xy^2\hat{x} + yz^2\hat{y} + 4x^2z\hat{z}$ and the surface below.

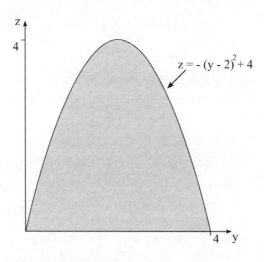
$z = -(y - 2)^2 + 4$

Solution The curl theorem states

$$\int_S (\nabla \times \vec{v}) \cdot d\vec{a} = \oint_\mathcal{P} \vec{v} \cdot d\vec{\ell}.$$

Starting with the left-hand side, the curl is given by

$$\nabla \times \vec{v} = \begin{vmatrix} \hat{x} & \hat{y} & \hat{z} \\ \dfrac{\partial}{\partial x} & \dfrac{\partial}{\partial y} & \dfrac{\partial}{\partial z} \\ 5xy^2 & yz^2 & 4x^2z \end{vmatrix} = -2yz\hat{x} - 8xz\hat{y} - 10xy\hat{z}$$

We also have $d\vec{a} = dy\, dz\, \hat{x}$ with $0 \leqslant z \leqslant -(y-2)^2 + 4$. So

$$\int_S (\nabla \times \vec{v}) \cdot d\vec{a} = \int_0^4 \int_0^{-(y-2)^2+4} -2yz\, dz\, dy = -\frac{1024}{15}.$$

Now to solve $\oint_\mathcal{P} \vec{v} \cdot d\vec{\ell}$ over the two paths (i) and (ii):

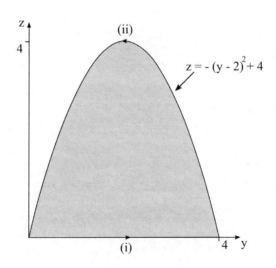

(i) Here we have $x = 0$, $z = 0$, and $d\vec{\ell} = dy\, \hat{y}$. So $\vec{v} \cdot d\vec{\ell} = y(0^2)dy = 0$.
(ii) Here we have $d\vec{\ell} = dy\, \hat{y} + dz\, \hat{z}$, $x = 0$, and $z = -(y-2)^2 + 4$

$$\int_\mathcal{P} \vec{v} \cdot d\vec{\ell} = \int_\mathcal{P} yz^2 dy + 4(0^2)z\, dz = \int_4^0 y\left[-(y-2)^2 + 4\right]^2 dy = -\frac{1024}{15}.$$

So,

$$\oint_P \vec{v} \cdot d\vec{\ell} = 0 + \frac{-1024}{15} = \frac{-1024}{15}$$

as expected.

Problem 1.12. Test the gradient theorem with $T = 3xz^2 - y^2z$ and path $z = y^2$ and $z = y^3$ from $(0, 0, 0) \to (0, 1, 1)$.

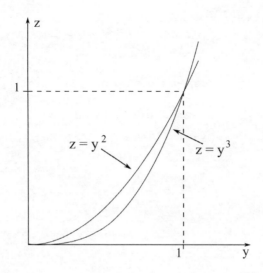

Solution The gradient theorem states

$$\int_{\vec{a}}^{\vec{b}} \nabla T \cdot d\vec{\ell} = T(\vec{b}) - T(\vec{a}).$$

Starting with the right side, we have

$$T(0,1,1) - T(0,0,0) = 3(0)(1^2) - (1^2)(1) - 0 = -1.$$

Now to solve $\int_{\vec{a}}^{\vec{b}} \nabla T \cdot d\vec{\ell}$, the gradient of T is given by

$$\nabla T = \frac{\partial}{\partial x}\left(3xz^2 - y^2z\right)\hat{x} + \frac{\partial}{\partial y}\left(3xz^2 - y^2z\right)\hat{y} + \frac{\partial}{\partial z}\left(3xz^2 - y^2z\right)\hat{z}$$

$$\nabla T = 3z^2\hat{x} - 2yz\,\hat{y} + \left(6xz - y^2\right)\hat{z}.$$

Here, $d\vec{\ell} = dy\,\hat{y} + dz\,\hat{z}$ with $x = 0$. So

$$\nabla T \cdot d\vec{\ell} = -2yz\,dy + \left[6(0)(z) - y^2\right]dz = -2yz\,dy - y^2 dz.$$

For path (i), we have $z = y^2 \to dz = 2y\,dy$. So

$$\nabla T \cdot d\vec{\ell} = -2y(y^2)dy - y^2(2y\,dy) = -4y^3\,dy$$

and

$$\int_{\vec{a}}^{\vec{b}} \nabla T \cdot d\vec{\ell} = \int_0^1 -4y^3 dy = -1$$

as expected. For path (ii), we have $z = y^3 \to dz = 3y^2 dy$. So

$$\nabla T \cdot d\vec{\ell} = -2y(y^3)dy - y^2(3y^2 dy) = -5y^4 dy$$

and

$$\int_{\vec{a}}^{\vec{b}} \nabla T \cdot d\vec{\ell} = \int_0^1 -5y^4 dy = -1$$

also as expected.

Problem 1.13. Verify the following integration by parts given $f = xy^2 z$ and $\vec{A} = z^2 \hat{x} + 4xy\hat{y} - x^2 z\hat{z}$ and the surface below,

$$\int_S f(\nabla \times \vec{A}) \cdot d\vec{a} = \int_S \left[\vec{A} \times (\nabla f)\right] \cdot d\vec{a} + \oint_P f\vec{A} \cdot d\vec{\ell}.$$

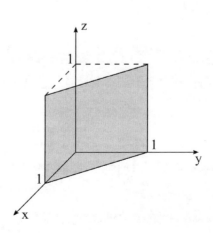

Solution Starting with the left-hand side

$$\nabla \times \vec{A} = \begin{vmatrix} \hat{x} & \hat{y} & \hat{z} \\ \frac{\partial}{\partial x} & \frac{\partial}{\partial y} & \frac{\partial}{\partial z} \\ z^2 & 4xy & -x^2z \end{vmatrix} = [2z - (-2xz)]\hat{y} + 4y\hat{z} = 2z(x+1)\hat{y} + 4y\hat{z}.$$

Now

$$f(\nabla \times \vec{A}) = 2xy^2z^2(x+1)\hat{y} + 4xy^3z\hat{z}.$$

Here we have $d\vec{a} = dx'\, dz\, \hat{n}$ where $\hat{n} = \hat{x} + \hat{y}$ and $n = \sqrt{2}$ so $\hat{n} = \frac{\sqrt{2}}{2}\hat{x} + \frac{\sqrt{2}}{2}\hat{y}$. Also from

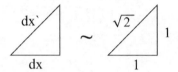

we have

$$dx' = \sqrt{2}\, dx \quad \text{with } y = 1 - x.$$

Now

$$d\vec{a} = \sqrt{2}\, dx\, dz \left(\frac{\sqrt{2}}{2}\hat{x} + \frac{\sqrt{2}}{2}\hat{y}\right) = dx\, dz(\hat{x} + \hat{y}).$$

Therefore,

$$\int_S f(\nabla \times \vec{A}) \cdot d\vec{a} = \int_0^1 \int_0^1 \left[2xy^2z^2(x+1)\hat{y} + 4xy^3z\hat{z}\right] \cdot (\hat{x} + \hat{y}) dx\, dz$$

$$= \int_0^1 \int_0^1 2x(1-x)^2 z^2(x+1) dx\, dz = \frac{7}{90}.$$

Next, we will solve the $\oint_P f\vec{A} \cdot d\vec{\ell}$ term for the four segments.

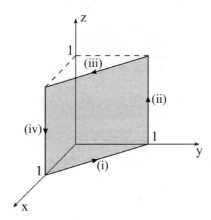

Segment (i)
$$z = 0 \rightarrow f = xy^2(0) = 0.$$

Segment (ii)
$$x = 0 \rightarrow f = (0)y^2 z = 0.$$

Segment (iii)
$$d\vec{\ell} = dx\,\hat{x} + dy\,\hat{y},\ z=1, \quad \text{and} \quad y = 1 - x \rightarrow dy = -dx.$$

Segment (iv)
$$y = 0 \rightarrow f = x(0^2)z = 0.$$

So
$$f(\vec{A}\cdot d\vec{\ell}) = xy^2(z^2 dx + 4xy\,dy) = x(1-x)^2\big[1 - 4x(1-x)\big]dx$$

and
$$\oint_{\mathcal{P}} f\vec{A}\cdot d\vec{\ell} = \int_0^1 x(1-x)^2\big[1 - 4x(1-x)\big]dx = \frac{1}{60}.$$

Now to solve the $\int_S [A \times \nabla f]\cdot d\vec{a}$ term. First, we have
$$\nabla f = y^2 z\,\hat{x} + 2xyz\,\hat{y} + xy^2\,\hat{z}.$$

So

$$A \times (\nabla f) = \begin{vmatrix} \hat{x} & \hat{y} & \hat{z} \\ z^2 & 4xy & -x^2z \\ y^2z & 2xyz & xy^2 \end{vmatrix}$$

$$= \left(4x^2y^3 + 2x^3yz^2\right)\hat{x} + \left(-x^2y^2z^2 - xy^2z^2\right)\hat{y} + \left(2xyz^3 - 4xy^3z\right)\hat{z}.$$

As before, $d\vec{a} = dx\, dz(\hat{x} + \hat{y})$. So

$$\int_S [A \times (\nabla f)] \cdot d\vec{a} = \int_0^1 \int_0^1 \Big[4x^2(1-x)^3 + 2x^3(1-x)z^2 \\ -x^2(1-x)^2z^2 - x(1-x)^2z^2\Big] dx\, dz$$

$$\int_S [A \times (\nabla f)] \cdot d\vec{a} = \frac{11}{180}.$$

So

$$\int_S [A \times (\nabla f)] \cdot d\vec{a} + \oint_P f\vec{A} \cdot d\vec{\ell} = \frac{11}{180} + \frac{1}{60} = \frac{7}{90}$$

as expected.

Problem 1.14. Find the divergence and curl of the following functions:
a) $\vec{v} = r^2 \hat{r} + \cos\theta \sin\phi\, \hat{\theta} + \sin\theta \cos\phi\, \hat{\phi}$
b) $\vec{v} = s\cos\phi\, \hat{s} + \cos\phi \sin\phi\, \hat{\phi} + z\sin\phi\, \hat{z}$

Solutions
a) $\vec{v} = r^2 \hat{r} + \cos\theta \sin\phi\, \hat{\theta} + \sin\theta \cos\phi\, \hat{\phi}$

$$\nabla \cdot \vec{v} = \frac{1}{r^2}\frac{\partial}{\partial r}(r^2 v_r) + \frac{1}{r\sin\theta}\frac{\partial}{\partial \theta}(\sin\theta\, v_\theta) + \frac{1}{r\sin\theta}\frac{\partial v_\phi}{\partial \phi}$$

$$= \frac{1}{r^2}\frac{\partial}{\partial r}(r^4) + \frac{1}{r\sin\theta}\frac{\partial}{\partial \theta}(\sin\theta \cos\theta \sin\phi) + \frac{1}{r\sin\theta}\frac{\partial}{\partial \phi}(\sin\theta \cos\phi)$$

$$= \frac{1}{r^2}(4r^3) + \frac{\sin\phi}{r\sin\theta}\left(-\sin^2\theta + \cos^2\theta\right) + \frac{1}{r}(-\sin\phi)$$

$$= 4r + \frac{\sin\phi}{r\sin\theta}(1 - 2\sin^2\theta) - \frac{\sin\phi}{r}$$

$$\nabla \cdot \vec{v} = 4r + \frac{\sin\phi}{r}(\csc\theta - 2\sin\theta - 1)$$

$$\nabla \times \vec{v} = \frac{1}{r\sin\theta}\left[\frac{\partial}{\partial\theta}(\sin\theta\, v_\phi) - \frac{\partial v_\theta}{\partial\phi}\right]\hat{r}$$

$$+ \frac{1}{r}\left[\frac{1}{\sin\theta}\frac{\partial v_r}{\partial\phi} - \frac{\partial}{\partial r}(rv_\phi)\right]\hat{\theta} + \frac{1}{r}\left[\frac{\partial}{\partial r}(rv_\theta) - \frac{\partial v_r}{\partial\theta}\right]\hat{\phi}$$

$$= \frac{1}{r\sin\theta}\left[\frac{\partial}{\partial\theta}(\sin^2\theta\cos\phi) - \frac{\partial}{\partial\phi}(\cos\theta\sin\phi)\right]\hat{r}$$

$$+ \frac{1}{r}\left[\frac{1}{\sin\theta}\frac{\partial}{\partial\phi}(r^2) - \frac{\partial}{\partial r}(r\sin\theta\cos\phi)\right]\hat{\theta}$$

$$+ \frac{1}{r}\left[\frac{\partial}{\partial r}(r\cos\theta\sin\phi) - \frac{\partial}{\partial\phi}(r^2)\right]\hat{\phi}$$

$$= \frac{1}{r\sin\theta}(2\sin\theta\cos\theta\cos\phi - \cos\theta\cos\phi)\hat{r} - \frac{\sin\theta\cos\phi}{r}\hat{\theta}$$

$$+ \frac{\cos\theta\sin\phi}{r}\hat{\phi}$$

$$\nabla \times \vec{v} = \frac{\cos\theta\cos\phi}{r}(2 - \csc\theta)\hat{r} - \frac{\sin\theta\cos\phi}{r}\hat{\theta} + \frac{\cos\theta\sin\phi}{r}\hat{\phi}$$

b) $\vec{v} = s\cos\phi\,\hat{s} + \cos\phi\sin\phi\,\hat{\phi} + z\sin\phi\,\hat{z}$

$$\nabla \cdot \vec{v} = \frac{1}{s}\frac{\partial}{\partial s}(sv_s) + \frac{1}{s}\frac{\partial v_\phi}{\partial\phi} + \frac{\partial v_z}{\partial z}$$

$$= \frac{1}{s}\frac{\partial}{\partial s}(s^2\cos\phi) + \frac{1}{s}\frac{\partial}{\partial\phi}(\cos\phi\sin\phi) + \frac{\partial}{\partial z}(z\sin\phi)$$

$$= 2\cos\phi + \frac{1}{s}(-\sin^2\phi + \cos^2\phi) + \sin\phi$$

$$\nabla \cdot \vec{v} = 2\cos\phi + \sin\phi + \frac{\cos^2\phi - \sin^2\phi}{s}$$

$$\nabla \times \vec{v} = \left(\frac{1}{s}\frac{\partial v_z}{\partial \phi} - \frac{\partial v_\phi}{\partial z}\right)\hat{s} + \left(\frac{\partial v_s}{\partial z} - \frac{\partial v_z}{\partial s}\right)\hat{\phi} + \frac{1}{s}\left[\frac{\partial}{\partial s}(sv_\phi) - \frac{\partial v_s}{\partial \phi}\right]\hat{z}$$

$$= \left[\frac{1}{s}\frac{\partial}{\partial \phi}(z\sin\phi) - \frac{\partial}{\partial z}(\cos\phi\sin\phi)\right]\hat{s} + \left[\frac{\partial}{\partial z}(s\cos\phi) - \frac{\partial}{\partial s}(z\sin\phi)\right]\hat{\phi}$$

$$+ \frac{1}{s}\left[\frac{\partial}{\partial s}(s\cos\phi\sin\phi) - \frac{\partial}{\partial \phi}(s\cos\phi)\right]\hat{z}$$

$$= \frac{z}{s}\cos\phi\,\hat{s} + \frac{1}{s}(\cos\phi\sin\phi + s\sin\phi)\hat{z}$$

$$\nabla \times \vec{v} = \frac{z}{s}\cos\phi\,\hat{s} + \frac{\sin\phi}{s}(\cos\phi + s)\hat{z}$$

Problem 1.15. Find the gradient and Laplacian of:
a) $T = r^2(\cos\theta\sin\phi + \sin\theta\cos\phi)$
b) $T = z^2\sin\phi - s\cos^2\phi$

Solutions
a) $T = r^2(\cos\theta\sin\phi + \sin\theta\cos\phi)$

$$\nabla T = \frac{\partial T}{\partial r}\hat{r} + \frac{1}{r}\frac{\partial T}{\partial \theta}\hat{\theta} + \frac{1}{r\sin\theta}\frac{\partial T}{\partial \phi}\hat{\phi}$$

$$= 2r(\cos\theta\sin\phi + \sin\theta\cos\phi)\hat{r} + \frac{1}{r}r^2(-\sin\theta\sin\phi + \cos\theta\cos\phi)\hat{\theta}$$

$$+ \frac{1}{r\sin\theta}r^2(\cos\theta\cos\phi - \sin\theta\sin\phi)\hat{\phi}$$

$$= 2r(\cos\theta\sin\phi + \sin\theta\cos\phi)\hat{r} + r(\cos\theta\cos\phi - \sin\theta\sin\phi)\hat{\theta}$$

$$+ \frac{r}{\sin\theta}(\cos\theta\cos\phi - \sin\theta\sin\phi)\hat{\phi}$$

$$\nabla T = 2r\sin(\theta + \phi)\hat{r} + r\cos(\theta + \phi)\hat{\theta} + \frac{r}{\sin\theta}\cos(\theta + \phi)\hat{\phi}.$$

Note we could have written T as $T = r^2\sin(\theta + \phi)$ and then computed the gradient.

$$\nabla^2 T = \frac{1}{r^2}\frac{\partial}{\partial r}\left(r^2\frac{\partial T}{\partial r}\right) + \frac{1}{r^2\sin\theta}\frac{\partial}{\partial \theta}\left(\sin\theta\frac{\partial T}{\partial \theta}\right) + \frac{1}{r^2\sin^2\theta}\left(\frac{\partial^2 T}{\partial \phi^2}\right)$$

$$= \frac{1}{r^2}\frac{\partial}{\partial r}\left[2r^3 \sin(\theta + \phi)\right] + \frac{1}{r^2 \sin\theta}\frac{\partial}{\partial \theta}\left[r^2 \sin\theta \cos(\theta + \phi)\right]$$
$$+ \frac{1}{r^2 \sin^2\theta}\frac{\partial}{\partial \phi}\left[r^2 \cos(\theta + \phi)\right]$$

$$= 6\sin(\theta + \phi) + \frac{1}{\sin\theta}[\cos\theta \cos(\theta + \phi) - \sin\theta \sin(\theta + \phi)]$$
$$+ \frac{1}{\sin^2\theta}(-\sin(\theta + \phi))$$

$$\nabla^2 T = 5\sin(\theta + \phi) + \frac{\cos\theta}{\sin\theta}\cos(\theta + \phi) - \frac{\sin(\theta + \phi)}{\sin^2\theta}.$$

b) $T = z^2 \sin\phi - s\cos^2\phi$

$$\nabla T = \frac{\partial T}{\partial s}\hat{s} + \frac{1}{s}\frac{\partial T}{\partial \phi}\hat{\phi} + \frac{\partial T}{\partial z}\hat{z}$$

$$= \frac{\partial}{\partial s}(z^2 \sin\phi - s\cos^2\phi)\hat{s} + \frac{1}{s}\frac{\partial}{\partial \phi}(z^2 \sin\phi - s\cos^2\phi)\hat{\phi}$$
$$+ \frac{\partial}{\partial z}(z^2 \sin\phi - s\cos^2\phi)\hat{z}$$

$$= -\cos^2\phi\,\hat{s} + \frac{1}{s}[z^2 \cos\phi - 2s\cos\phi(-\sin\phi)]\hat{\phi} + 2z\sin\phi\,\hat{z}$$

$$\nabla T = -\cos^2\phi\,\hat{s} + \frac{\cos\phi}{s}(z^2 + 2s\sin\phi)\hat{\phi} + 2z\sin\phi\,\hat{z}$$

$$\nabla^2 T = \frac{1}{s}\frac{\partial}{\partial s}\left(s\frac{\partial T}{\partial s}\right) + \frac{1}{s^2}\frac{\partial^2 T}{\partial \phi^2} + \frac{\partial^2 T}{\partial z^2}$$

$$\frac{\partial T}{\partial s} = -\cos^2\phi \to s\frac{\partial T}{\partial s} = -s\cos^2\phi \to \frac{\partial}{\partial s}\left(s\frac{\partial T}{\partial s}\right) = -\cos^2\phi$$

$$\frac{\partial T}{\partial \phi} = z^2 \cos\phi + 2s\cos\phi\sin\phi \to \frac{\partial^2 T}{\partial \phi^2} = -z^2 \sin\phi + 2s(-\sin^2\phi + \cos^2\phi)$$

$$\frac{\partial T}{\partial z} = 2z\sin\phi \to \frac{\partial^2 T}{\partial z^2} = 2\sin\phi$$

$$\nabla^2 T = -\frac{\cos^2\phi}{s} + \frac{2}{s}(-\sin^2\phi + \cos^2\phi) - \frac{z^2}{s^2}\sin\phi + 2\sin\phi$$

$$\nabla^2 T = \frac{\cos^2 \phi}{s} - \frac{2}{s} \sin^2 \phi + \left(2 - \frac{z^2}{s^2}\right) \sin \phi$$

Problem 1.16. Test the divergence theorem with $\vec{v} = r \cos \phi \, \hat{r} + r \cos \theta \sin \theta \, \hat{\theta} + r \sin \phi \, \hat{\phi}$ and the volume below (the upper half of the sphere of radius R with a cone of radius $a = \frac{R}{\sqrt{3}}$ cut out).

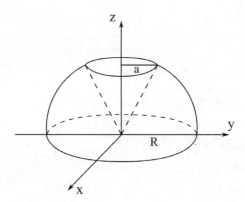

Solution The divergence theorem states

$$\int_v \nabla \cdot \vec{v} \, d\tau = \oint_S \vec{v} \cdot d\vec{a}.$$

Starting with the left-hand side, the divergence is

$$\nabla \cdot \vec{v} = \frac{1}{r^2} \frac{\partial}{\partial r} (r^2 v_r) + \frac{1}{r \sin \theta} \frac{\partial}{\partial \theta} (\sin \theta \, v_\theta) + \frac{1}{r \sin \theta} \frac{\partial v_\phi}{\partial \phi}$$

$$= \frac{1}{r^2} \frac{\partial}{\partial r} (r^3 \cos \phi) + \frac{1}{r \sin \theta} \frac{\partial}{\partial \theta} (r \sin^2 \theta \cos \theta) + \frac{1}{r \sin \theta} \frac{\partial}{\partial \phi} (r \sin \phi)$$

$$= 3 \cos \phi + \frac{1}{\sin \theta} (2 \sin \theta \cos^2 \theta - \sin^3 \theta) + \frac{\cos \phi}{\sin \theta}$$

$$\nabla \cdot \vec{v} = 3 \cos \phi + 2 \cos^2 \theta - \sin^2 \theta + \frac{\cos \phi}{\sin \theta}.$$

For the volume,

$$0 \leq r \leq R, \quad \tan^{-1}\left(\frac{a}{R}\right) = \tan^{-1}\left(\frac{1}{\sqrt{3}}\right) = \frac{\pi}{6} \to \frac{\pi}{6} \leq \theta \leq \frac{\pi}{2}, \quad 0 \leq \phi \leq 2\pi.$$

So

$$\int_V \vec{\nabla} \cdot \vec{v} \, d\tau = \int_0^R \int_{\frac{\pi}{6}}^{\frac{\pi}{2}} \int_0^{2\pi} \left(3\cos\phi + 2\cos^2\theta - \sin^2\theta + \frac{\cos\phi}{\sin\theta} \right) (r^2 \sin\theta) d\phi \, d\theta \, dr$$

$$\int_V \vec{\nabla} \cdot \vec{v} \, d\tau = -\frac{\sqrt{3}}{12} \pi R^3.$$

Now for the right-hand side, we have three surfaces: the bottom (i), the outer shell (ii), and the inner part where the cone is cut out (iii). We have

$$\vec{v} = r\cos\phi \, \hat{r} + r\sin\phi \, \hat{\phi} + r\cos\theta \sin\theta \, \hat{\theta}.$$

For (i), we have $d\vec{a} = r \, dr \, d\phi \, \hat{\theta}$ and $\theta = \frac{\pi}{2}$. So

$$\vec{v} \cdot d\vec{a} = r^2 \cos\frac{\pi}{2} \sin\frac{\pi}{2} dr \, d\phi = 0$$

and

$$\int_{(i)} \vec{v} \cdot d\vec{a} = 0.$$

For (ii), we have $r = R$ and $d\vec{a} = r^2 \sin\theta \, d\theta \, d\phi \, \hat{r} = R^2 \sin\theta \, d\theta \, d\phi \, \hat{r}$. So

$$\vec{v} \cdot d\vec{a} = R^3 \cos\phi \sin\theta \, d\theta \, d\phi$$

and

$$\int_{(ii)} \vec{v} \cdot d\vec{a} = R^3 \int_0^{2\pi} \int_{\frac{\pi}{6}}^{\frac{\pi}{2}} \cos\phi \sin\theta \, d\theta \, d\phi = 0.$$

For (iii), we have $\theta = \frac{\pi}{6}$ and $d\vec{a} = -r\sin\theta \, dr \, d\phi \, \hat{\theta} = -\frac{1}{2} r \, dr \, d\phi \, \hat{\theta}$. So

$$\vec{v} \cdot d\vec{a} = -\frac{1}{2} r^2 \cos\frac{\pi}{6} \sin\frac{\pi}{6} = -\frac{\sqrt{3}}{8} r^2$$

and

$$\int_{(iii)} \vec{v} \cdot d\vec{a} = -\frac{\sqrt{3}}{8} \int_0^R \int_0^{2\pi} r^2 d\phi \, dr = -\frac{\sqrt{3}}{12} \pi R^3.$$

Therefore,
$$\oint_S \vec{v} \cdot d\vec{a} = -\frac{\sqrt{3}}{12}\pi R^3$$
as expected.

Problem 1.17. Test the curl theorem with $\vec{v} = s^2 z\,\hat{s} + \sin\phi\cos\phi\,\hat{\phi} + zs\cos\phi\,\hat{z}$ and half of a cylindrical shell with radius R and height h.

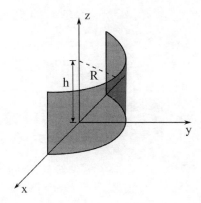

Solution The curl theorem states
$$\int_S \left(\nabla \times \vec{v}\right) \cdot d\vec{a} = \oint_P \vec{v} \cdot d\vec{\ell}.$$
Starting with the left-handed side, we have
$$d\vec{a} = s\,d\phi\,dz\,\hat{s} = R\,d\phi\,dz\,\hat{s}.$$
Since we are dotting $d\vec{a}$ with $\nabla \times \vec{v}$, we only need the $\hat{s}$ component of the curl:
$$[\nabla \times \vec{v}]_s = \left(\frac{1}{s}\frac{\partial v_z}{\partial \phi} - \frac{\partial v_\phi}{\partial z}\right)\hat{s} = \left[\frac{1}{s}\frac{\partial}{\partial \phi}(zs\cos\phi) - \frac{\partial}{\partial z}(\sin\phi\cos\phi)\right]\hat{s}$$
$$= -z\sin\phi\,\hat{s}.$$
So
$$\left(\nabla \times \vec{v}\right) \cdot d\vec{a} = -Rz\sin\phi\,d\phi\,dz.$$

We have
$$0 \leqslant \phi \leqslant \pi \quad \text{and} \quad 0 \leqslant z \leqslant h$$
so
$$\int_S \left(\vec{\nabla} \times \vec{v}\right) \cdot d\vec{a} = \int_0^h \int_0^\pi -Rz \sin\phi \, d\phi \, dz = -h^2 R.$$

For the left-hand side, we have four curves

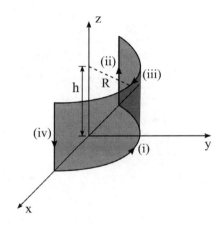

with
$$\vec{v} = s^2 z \, \hat{s} + \sin\phi \cos\phi \, \hat{\phi} + z \cos\phi \, \hat{z}.$$

For curve (i), $d\vec{\ell} = d\phi \, \hat{\phi}$, $z = 0$, and $s = R$. So
$$\vec{v} \cdot d\vec{\ell} = \sin\phi \cos\phi \, d\phi$$
and
$$\int_0^\pi \sin\phi \cos\phi \, d\phi = 0.$$

For curve (ii), $d\vec{\ell} = dz \, \hat{z}$, $\phi = \pi$, and $s = R$. So
$$\vec{v} \cdot d\vec{\ell} = zs \cos\phi \, dz = zR \cos\pi \, dz = -zR \, dz$$
and
$$\int_0^h -zR \, dz = -\frac{1}{2} h^2 R.$$

For curve (iii), $d\vec{\ell} = d\phi\,\hat{\phi}$, $z = h$, and $s = R$. So
$$\vec{v}\cdot d\vec{\ell} = \sin\phi\cos\phi\,d\phi$$
and
$$\int_{\pi}^{0}\sin\phi\cos\phi\,d\phi = 0.$$

For curve (iv), $d\vec{\ell} = dz\,\hat{z}$, $\phi = 0$, and $s = R$. So
$$\vec{v}\cdot d\vec{\ell} = zs\cos\phi\,dz = zR\cos(0)dz = zR\,dz$$
and
$$\int_{h}^{0} zR\,dz = -\frac{1}{2}h^2 R^2.$$

So,
$$\oint_{\mathcal{P}} \vec{v}\cdot d\vec{\ell} = -\frac{1}{2}h^2 R - \frac{1}{2}h^2 R = -h^2 R$$

as expected.

Problem 1.18. Test the gradient theorem using $T = sz^2\sin\phi$ and the half helix path (radius R, height h).

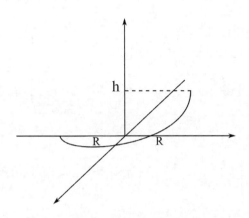

Solution The gradient theorem states
$$\int_{\mathcal{P}} \nabla T \cdot d\vec{\ell} = T(\vec{b}) - T(\vec{a}).$$

Starting with the right-hand side
$$T(\vec{b}) - T(\vec{a}) = T\left(R, \frac{\pi}{2}, h\right) - T\left(R, -\frac{\pi}{2}, 0\right) = Rh^2 \sin\frac{\pi}{2} - R(0)^2 \sin\left(-\frac{\pi}{2}\right) = h^2 R.$$

Now, the gradient is
$$\nabla T = \frac{\partial T}{\partial s}\hat{s} + \frac{1}{s}\frac{\partial T}{\partial \phi}\hat{\phi} + \frac{\partial T}{\partial z}\hat{z} = z^2 \sin\phi \, \hat{s} + z^2 \cos\phi \, \hat{\phi} + 2sz \sin\phi \, \hat{z}.$$

We also have $s = R$ and $\vec{\ell} = s \, d\phi \, \hat{\phi} + dz \, \hat{z} = R \, d\phi \, \hat{\phi} + dz \, \hat{z}$. So
$$\nabla T \cdot d\vec{\ell} = Rz^2 \cos\phi \, d\phi + 2Rz \sin\phi \, dz.$$

We need a way to relate z and ϕ. Note that as ϕ increases, z increases linearly. So, using the equation of line
$$z - z_0 = \gamma(\phi - \phi_0),$$

when $z = 0$ and $\phi = -\frac{\pi}{2}$,
$$z = \gamma\left(\phi + \frac{\pi}{2}\right),$$

when $z = h$ and $\phi = \frac{\pi}{2}$,
$$h = \gamma\left(\frac{\pi}{2} + \frac{\pi}{2}\right) \to \gamma = \frac{h}{\pi},$$

so
$$z = \frac{h}{\pi}\phi - \frac{h}{2}$$

and
$$dz = \frac{h}{\pi}d\phi.$$

Using our expressions for z and dz, we have
$$\nabla T \cdot d\vec{\ell} = \left[R\left(\frac{h}{\pi}\phi + \frac{h}{2}\right)^2 \cos\phi + 2R\left(\frac{h}{\pi}\phi + \frac{h}{2}\right) \sin\phi \left(\frac{h}{\pi}\right)\right]d\phi.$$

So
$$\int_{\vec{a}}^{\vec{b}} \nabla T \cdot d\vec{\ell} = \int_{-\frac{\pi}{2}}^{\frac{\pi}{2}} \left[R\left(\frac{h}{\pi}\phi + \frac{h}{2}\right)^2 \cos\phi + 2R\left(\frac{h}{\pi}\phi + \frac{h}{2}\right) \sin\phi \left(\frac{h}{\pi}\right)\right]d\phi = h^2 R$$

as expected.

Problem 1.19. Evaluate the following integrals:

a) $\int_{1}^{3} (2x^2 - x + 4)\delta(x - 2)dx$

b) $\int_{-1}^{1} (x^2 + 4)\delta(x - 2)dx$

c) $\int_{2}^{6} \sin(\frac{3x}{2})\delta(x - \pi)dx$

d) $\int_{-2}^{2} (2x^3 + 1)\delta(4x)dx$

e) $\int_{-\infty}^{\infty} x^2 \delta(2x + 1)dx$

f) $\int_{0}^{a} \delta(x - b)dx$

Solutions

a)
$$\int_{1}^{3} (2x^2 - x + 4)\delta(x - 2)dx.$$

Since $2 \in (1, 3)$ and $f(x) = 2x^2 - x + 4$, we have
$$\int_{1}^{3} (2x^2 - x + 4)\delta(x - 2)dx = f(2) = 2(2)^2 - 2 + 4 = 10.$$

b)
$$\int_{-1}^{1} (x^2 + 4)\delta(x - 2)dx.$$

Since $2 \notin (-1, 1)$, we have
$$\int_{-1}^{1} (x^2 + 4)\delta(x - 2)dx = 0.$$

c)
$$\int_{2}^{6} \sin\left(\frac{3x}{2}\right)\delta(x - \pi)dx.$$

Since $\pi \in (2, 6)$ and $f(x) = \sin(\frac{3x}{2})$, we have

$$\int_2^6 \sin\left(\frac{3x}{2}\right)\delta(x - \pi)dx = f(\pi) = \sin\left(\frac{3\pi}{2}\right) = -1.$$

d)
$$\int_{-2}^2 (2x^3 + 1)\delta(4x)dx.$$

Since $0 \in (-2, 2)$ and $f(x) = 2x^3 + 1$, we have

$$\int_{-2}^2 (2x^3 + 1)\delta(4x)dx = \frac{1}{|4|}f(0) = \frac{1}{|4|}(2(0)^3 + 1) = \frac{1}{4}.$$

e)
$$\int_{-\infty}^{\infty} x^2\delta(2x + 1)dx.$$

This can be rewritten as

$$\int_{-\infty}^{\infty} x^2\delta(2x + 1)dx = \int_{-\infty}^{\infty} x^2\delta\left[2\left(x + \frac{1}{2}\right)\right]dx = \int_{-\infty}^{\infty} x^2\frac{1}{|2|}\delta\left(x + \frac{1}{2}\right)dx.$$

Since $-\frac{1}{2} \in (-\infty, \infty)$ and $f(x) = x^2$, we have

$$\int_{-\infty}^{\infty} x^2\frac{1}{|2|}\delta\left(x + \frac{1}{2}\right)dx = \frac{1}{|2|}f\left(-\frac{1}{2}\right) = \frac{1}{|2|}\left(-\frac{1}{2}\right)^2 = \frac{1}{8}.$$

f)
$$\int_0^a \delta(x - b)dx.$$

Here we have

$$\int_0^a \delta(x - b)dx = \begin{cases} 1 & \text{if } 0 < b < a \\ 0 & \text{otherwise} \end{cases}.$$

Problem 1.20. Suppose we have two vector fields $\vec{F}_1 = y^2\hat{z}$ and $\vec{F}_2 = x\hat{x} + y\hat{y} + z\hat{z}$. Calculate the divergence and curl of each. Which can be written as the gradient of a scalar and which can be written as the curl of a vector? Find a scalar and a vector potential.

Solution For $\vec{F}_1$, we have

$$\nabla \cdot \vec{F}_1 = \left(\frac{\partial}{\partial x}\hat{x} + \frac{\partial}{\partial y}\hat{y} + \frac{\partial}{\partial z}\hat{z}\right) \cdot (y^2\hat{z}) = \frac{\partial(y^2)}{\partial z} = 0$$

and

$$\nabla \times \vec{F}_1 = \begin{vmatrix} \hat{x} & \hat{y} & \hat{z} \\ \frac{\partial}{\partial x} & \frac{\partial}{\partial y} & \frac{\partial}{\partial z} \\ 0 & 0 & y^2 \end{vmatrix} = 2y\hat{x}.$$

For $\vec{F}_2$, we have

$$\nabla \cdot \vec{F}_2 = \left(\frac{\partial}{\partial x}\hat{x} + \frac{\partial}{\partial y}\hat{y} + \frac{\partial}{\partial z}\hat{z}\right) \cdot (x\hat{x} + y\hat{y} + z\hat{z}) = 1 + 1 + 1 = 3$$

and

$$\nabla \times \vec{F}_2 = \begin{vmatrix} \hat{x} & \hat{y} & \hat{z} \\ \frac{\partial}{\partial x} & \frac{\partial}{\partial y} & \frac{\partial}{\partial z} \\ x & y & z \end{vmatrix} = (0-0)\hat{x} + (0-0)\hat{y} + (0-0)\hat{z} = 0.$$

Since $\nabla \cdot \vec{F}_1 = 0$, $\vec{F}_1$ can be expressed as $\vec{F}_1 = \nabla \times \vec{A}$. We can find $\vec{A}$ by considering

$$\nabla \times \vec{A} = \begin{vmatrix} \hat{x} & \hat{y} & \hat{z} \\ \frac{\partial}{\partial x} & \frac{\partial}{\partial y} & \frac{\partial}{\partial z} \\ 0 & 0 & y^2 \end{vmatrix}$$

$$= \left(\frac{\partial A_z}{\partial y} - \frac{\partial A_y}{\partial z}\right)\hat{x} + \left(\frac{\partial A_x}{\partial z} - \frac{\partial A_z}{\partial x}\right)\hat{y} + \left(\frac{\partial A_y}{\partial x} - \frac{\partial A_x}{\partial y}\right)\hat{z}.$$

By inspection:

$$\frac{\partial A_z}{\partial y} - \frac{\partial A_y}{\partial z} = 0, \quad \frac{\partial A_x}{\partial z} - \frac{\partial A_z}{\partial x} = 0, \quad \frac{\partial A_y}{\partial x} - \frac{\partial A_x}{\partial y} = y^2.$$

This is satisfied by

$$\vec{A} = y^2 x \hat{y},$$

which is just one example. Since $\nabla \times \vec{F}_2 = 0$, $\vec{F}_2$ can be expressed as $\vec{F}_2 = -\nabla V$. We can find V by considering

$$\vec{F}_2 = -\left(\frac{\partial V}{\partial x}\hat{x} + \frac{\partial V}{\partial y}\hat{y} + \frac{\partial V}{\partial z}\hat{z}\right).$$

By inspection:

$$x = -\frac{\partial V}{\partial x}, \qquad y = -\frac{\partial V}{\partial y}, \qquad z = -\frac{\partial V}{\partial z}.$$

This is satisfied by

$$V = -\left(\frac{x^2}{2} + \frac{y^2}{2} + \frac{z^2}{2}\right)$$

which is again just one example.

Bibliography

Byron F W and Fuller R W 1992 *Mathematics of Classical and Quantum Physics* (New York: Dover)
Griffiths D J 1999 *Introduction to Electrodynamics* 3rd edn (Englewood Cliffs, NJ: Prentice Hall)
Griffiths D J 2013 *Introduction to Electrodynamics* 4th edn (New York: Pearson)
Halliday D, Resnick R and Walker J 2010 *Fundamentals of Physics* 9th edn (New York: Wiley)
Halliday D, Resnick R and Walker J 2013 *Fundamentals of Physics* 10th edn (New York: Wiley)
Purcell E M and Morin D J 2013 *Electricity and Magnetism* 3rd edn (Cambridge: Cambridge University Press)
Rogawski J 2011 *Calculus: Early Transcendentals* 2nd edn (San Francisco, CA: Freeman)

IOP Concise Physics

Electromagnetism
Problems and solutions
Carolina C Ilie and Zachariah S Schrecengost

Chapter 2

Electrostatics

Electrostatics is the topic of this chapter. Coulomb's law, Gauss's law, and the energy of various charge distributions are a few ways of understanding the electric field. The methods employed will make use of the specific degrees of symmetry. The mathematical skills obtained in chapter 1 will be applied here to analyze different charge distributions in Cartesian, spherical, or cylindrical coordinates.

2.1 Theory

2.1.1 Coulomb's law

The force on a point charge q due to a charge Q, separated by a distance r, is given by

$$\vec{F} = \frac{1}{4\pi\varepsilon_o} q \frac{Q}{r^2} \hat{r},$$

where $\varepsilon_0 (= 8.85 \times 10^{-12} \frac{C^2}{Nm^2})$ is the permittivity of free space.

2.1.2 Electric field

In general, for a volume charge density $\rho(\vec{r})$, the electric field at $\vec{r}$ is given by

$$\vec{E}(\vec{r}) = \frac{1}{4\pi\varepsilon_o} \int_v \frac{\rho(\vec{r}')}{r^2} \hat{r} \, d\tau'.$$

For a surface charge density $\sigma(\vec{r})$, the electric field is given by

$$\vec{E}(\vec{r}) = \frac{1}{4\pi\varepsilon_o} \int_s \frac{\sigma(\vec{r}')}{r^2} \hat{r} \, da'.$$

For a linear charge density $\lambda(\vec{r})$, the electric field is given by

$$\vec{E}(\vec{r}) = \frac{1}{4\pi\varepsilon_o} \int_\mathcal{P} \frac{\lambda(\vec{r}')}{r^2} \hat{r} \, d\ell'.$$

2.1.3 Gauss's law

For an electric field $\vec{E}$ and surface $\mathcal{S}$, Gauss's law states

$$\oint_\mathcal{S} \vec{E} \cdot d\vec{a} = \frac{q_{enc}}{\varepsilon_0},$$

where the enclosed charge is

$$q_{enc} = \int_\mathcal{V} \rho \, d\tau.$$

In differential form

$$\nabla \cdot \vec{E} = \frac{\rho}{\varepsilon_o},$$

where ρ is the volume charge density.

2.1.4 Curl of $\vec{E}$

$$\oint_\mathcal{P} \vec{E} \cdot d\vec{\ell} = 0 \;\rightarrow\; \nabla \times \vec{E} = 0,$$

where $\vec{E}$ is an *electrostatic* field.

2.1.5 Energy of a point charge distribution

The energy required to assemble n charges $q_1, q_2, \ldots, q_n$ is given by

$$W = \frac{1}{2} \sum_{i=i}^{n} q_i \left(\sum_{\substack{j=i \\ j \neq i}}^{n} \frac{1}{4\pi\varepsilon_o} \frac{q_j}{r_{ij}} \right) = \frac{1}{2} \sum_{i=1}^{n} q_i V(\vec{r}_i),$$

where $V(\vec{r}_i)$ is the potential at charge q_i and r_{ij} is the distance between charges q_i and q_j.

2.1.6 Energy of a continuous distribution

$$W = \frac{1}{2} \int \rho V(\vec{r}) \, d\tau = \frac{\varepsilon_o}{2} \int_{\text{all space}} E^2 \, d\tau.$$

2.1.7 Energy per unit volume

$$\frac{W}{\text{volume}} = \frac{\varepsilon_o}{2} E^2$$

2.2 Problems and solutions

Problem 2.1. Given the charge distribution below, find the force on charge $q_1 = q$ with $q_2 = 3q$, $q_3 = -2q$, and $q_4 = q$.

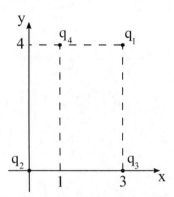

Solution The force on q_1 from q_2 is given by

$$\vec{F}_{21} = \frac{1}{4\pi\varepsilon_o} \frac{q_1 q_2}{r^2} \hat{r} = \frac{1}{4\pi\varepsilon_o} \frac{q_1 q_2}{r^3} \vec{r},$$

where

$$\vec{r} = 3\hat{x} + 4\hat{y}$$

and

$$r = \sqrt{3^2 + 4^2} = 5.$$

So

$$\vec{F}_{21} = \frac{1}{4\pi\varepsilon_o} \frac{3q^2}{5^3}(3\hat{x} + 4\hat{y}) = \frac{q^2}{4\pi\varepsilon_o}\left(\frac{9}{125}\hat{x} + \frac{12}{125}\hat{y}\right).$$

The force on q_1 from q_3 is given by

$$\vec{F}_{31} = \frac{1}{4\pi\varepsilon_o} \frac{q_1 q_3}{r^2} \hat{r},$$

where

$$r = 4$$

and

$$\hat{r} = \hat{y}.$$

So
$$\vec{F}_{31} = \frac{1}{4\pi\varepsilon_o}\left(-\frac{2q^2}{4^2}\right)\hat{y} = -\frac{q^2}{4\pi\varepsilon_o}\frac{1}{8}\hat{y}.$$

The force on q_1 from q_4 is given by
$$\vec{F}_{41} = \frac{1}{4\pi\varepsilon_o}\frac{q_1 q_4}{r^2}\hat{r},$$

where
$$r = 2$$

and
$$\hat{r} = \hat{x}.$$

So
$$\vec{F}_{41} = \frac{1}{4\pi\varepsilon_o}\frac{q^2}{2^2}\hat{r} = \frac{q^2}{4\pi\varepsilon_o}\frac{1}{4}\hat{x}.$$

Therefore,
$$\vec{F}_1 = \vec{F}_{21} + \vec{F}_{31} + \vec{F}_{41} = \frac{q^2}{4\pi\varepsilon_o}\left[\left(\frac{9}{125} + \frac{1}{4}\right)\hat{x} + \left(\frac{12}{125} - \frac{1}{8}\right)\hat{y}\right]$$
$$\vec{F}_1 = \frac{q^2}{4\pi\varepsilon_o}\left(\frac{161}{500}\hat{x} - \frac{29}{1000}\hat{y}\right).$$

Problem 2.2. Given a charged sheet with surface charge density $\sigma = ky$ (where k is a constant) and sides of length $2d$, find the electric field z above the center of the sheet.

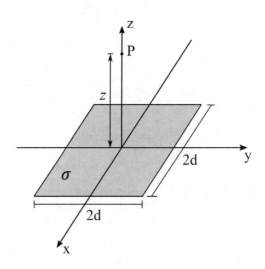

The electric field is given by
$$\vec{E} = \frac{1}{4\pi\varepsilon_o} \int \frac{\sigma}{r^2} \hat{r}\, da.$$

The horizontal components cancel so we only have the $\hat{z}$-component:
$$\hat{r} \to \cos\theta\, \hat{z} = \frac{z}{r}\hat{z}.$$

Also we have $da = dx\, dy$ and $r^2 = x^2 + y^2 + z^2$. Note that the piece of the sheet in each quadrant of the xy-plane contributes the same amount to the total field. Therefore,

$$\vec{E} = 4\vec{E}_{\text{quad}} = \frac{4}{4\pi\varepsilon_o} \int_0^d\int_0^d \frac{kyz\hat{z}}{(x^2+y^2+z^2)^{3/2}}dx\,dy = \frac{kz\hat{z}}{\pi\varepsilon_o}\int_0^d\int_0^d \frac{y}{(x^2+y^2+z^2)^{3/2}}dx\,dy$$

$$= \frac{kz\hat{z}}{\pi\varepsilon_o}\int_0^d y\left[\frac{x}{(y^2+z^2)\sqrt{y^2+z^2+x^2}}\right]_{x=0}^{x=d} dy = \frac{kzd\hat{z}}{\pi\varepsilon_o}\int_0^d \frac{y}{(y^2+z^2)\sqrt{y^2+z^2+d^2}}dy.$$

Let
$$u^2 = y^2 + z^2$$
so
$$2u\,du = 2y\,dy \to u\,du = y\,dy.$$

Evaluating u at the endpoints yields
$$u^2(y=0) = z^2 \to u = z$$
$$u^2(y=d) = d^2 + z^2 \to u = \sqrt{d^2+z^2}.$$

Now
$$\vec{E} = \frac{kzd\hat{z}}{\pi\varepsilon_o}\int_z^{\sqrt{z^2+d^2}} \frac{du}{u\sqrt{u^2+d^2}}$$

$$= \frac{kzd\hat{z}}{\pi\varepsilon_o}\left[\frac{1}{d}\ln\left(\frac{d+\sqrt{d^2+u^2}}{u}\right)\right]_{u=\sqrt{z^2+d^2}}^{u=z} = \frac{kz\hat{z}}{\pi\varepsilon_o}\ln\left(\frac{\frac{d+\sqrt{d^2+z^2}}{z}}{\frac{d+\sqrt{2d^2+z^2}}{\sqrt{z^2+d^2}}}\right)$$

$$\vec{E} = \frac{kz}{\pi\varepsilon_o}\ln\left[\frac{(d+\sqrt{z^2+d^2})\sqrt{z^2+d^2}}{z(d+\sqrt{2d^2+z^2})}\right]\hat{z}.$$

2-5

Problem 2.3. Find the electric field d above a cylinder of radius R, height h, and volume density ρ (ignoring edge effects).

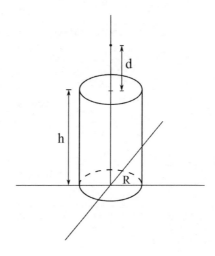

Solution We have

$$\vec{E} = \frac{1}{4\pi\varepsilon_o} \int \frac{\rho}{r^2} \hat{r} \, d\tau.$$

Note our horizontal components cancel, so $\hat{r} \to \cos\theta \, \hat{z}$ with

$$\cos\theta = \frac{d+h-z}{r}.$$

Also

$$d\tau = s \, ds \, d\phi \, dz$$

and

$$r^2 = s^2 + (d+h-z)^2.$$

Therefore,

$$\vec{E} = \frac{\rho}{4\pi\varepsilon_o} \int_0^{2\pi} \int_0^R \int_0^h \frac{(d+h-z)s \, \hat{z}}{\left[s^2 + (d+h-z)^2\right]^{3/2}} dz \, ds \, d\phi$$

$$= \frac{2\pi\rho}{4\pi\varepsilon_o} \left[\sqrt{R^2+d^2} - \sqrt{R^2+(d+h)^2} + h\right]\hat{z}$$

$$\vec{E} = \frac{\rho}{2\varepsilon_o} \left[\sqrt{R^2+d^2} - \sqrt{R^2+(d+h)^2} + h\right]\hat{z}.$$

Note if $R \gg d$ and $R \gg h$, the field reduces to

$$\vec{E} = \frac{\rho h}{2\varepsilon_o}\hat{z},$$

which is the field given by an infinite sheet of surface charge $\sigma = \rho h$.

Problem 2.4. Given the bottom hemisphere of a spherical shell of radius R, thickness d, and volume charge density ρ, find the electric field z above the center (above the open part, ignoring edge effects).

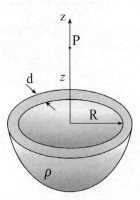

Solution The electric field is given by

$$\vec{E} = \frac{1}{4\pi\varepsilon_o} \int \frac{\rho}{r^2} \hat{r} \, d\tau.$$

We can see from

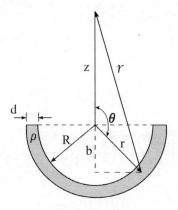

that

$$R \leqslant r \leqslant R + d, \quad 0 \leqslant \phi \leqslant 2\pi, \quad \frac{\pi}{2} \leqslant \theta \leqslant \pi.$$

Also, $d\tau = r^2 \sin\theta \, dr \, d\phi \, d\theta$. From the law of cosines,
$$\mathcal{r}^2 = z^2 + r^2 - 2rz\cos\theta.$$
Since the horizontal components cancel, $\hat{\mathcal{r}}$ becomes
$$\hat{\mathcal{r}} \to \cos\gamma \, \hat{z} = \frac{z+b}{\mathcal{r}}\hat{z},$$
where b is given by
$$\cos(\pi - \theta) = \frac{b}{r} \to b = -r\cos\theta.$$
So
$$\hat{\mathcal{r}} \to \frac{z - r\cos\theta}{\mathcal{r}}\hat{z}.$$
Therefore,
$$\vec{E} = \frac{\rho}{4\pi\varepsilon_o} \int_R^{R+d} \int_{\frac{\pi}{2}}^{\pi} \int_0^{2\pi} \frac{r^2(z - r\cos\theta)\sin\theta \, \hat{z}}{(z^2 + r^2 - 2rz\cos\theta)^{3/2}} d\phi \, d\theta \, dr$$

$$= \frac{\rho\hat{z}}{2\varepsilon_o} \int_R^{R+d} \int_{\frac{\pi}{2}}^{\pi} \frac{r^2(z - r\cos\theta)\sin\theta \, \hat{z}}{(z^2 + r^2 - 2rz\cos\theta)^{3/2}} d\theta \, dr.$$

Let $u = \cos\theta$ and $du = -\sin\theta \, d\theta$

$$u\left(\theta = \frac{\pi}{2}\right) = 0$$

$$u(\theta = \pi) = -1$$

$$\vec{E} = \frac{\rho\hat{z}}{2\varepsilon_o} \int_R^{R+d} \int_{-1}^{0} \frac{r^2(z - ru)\hat{z}}{(z^2 + r^2 - 2rzu)^{3/2}} du \, dr$$

$$= \frac{\rho\hat{z}}{2\varepsilon_o z^2} \int_R^{R+d} \left(r^2 - \frac{r^3}{\sqrt{r^2 + z^2}}\right) dr$$

$$= \frac{\rho\hat{z}}{2\varepsilon_o z^2} \left[\frac{r^3}{3} - \frac{\sqrt{r^2 + z^2}(r^2 - 2z^2)}{3}\right]_{r=R}^{r=R+d}$$

$$\vec{E} = \frac{\rho\hat{z}}{6\varepsilon_o z^2} \left\{(R+d)^3 - \sqrt{(R+d)^2 + z^2}\left[(R+d)^2 - 2z^2\right]\right.$$
$$\left. - R^3 + \sqrt{(R^2 + z^2)}(R^2 - 2z^2)\right\}.$$

Problem 2.5. Given the electric field $\vec{E} = k[2xz\hat{x} + z^2\hat{y} + (x^2 + 2yz)\hat{z}]$ (with constant k) find the following:
a) The charge density ρ.
b) The charge enclosed by a cylinder of height h, radius R, and base on the xy-plane center at the origin (below).
c) The charge enclosed by an upper hemisphere of radius R centered at the origin.

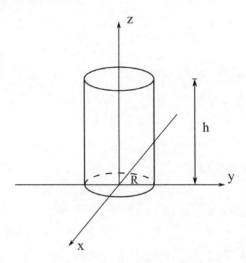

Solutions
a) The charge density ρ.
Gauss's law states

$$\nabla \cdot \vec{E} = \frac{\rho}{\varepsilon_o}.$$

So

$$\rho = \varepsilon_o \nabla \cdot \vec{E} = k\varepsilon_o(2y + 2z) = 2k\varepsilon_o(y + z).$$

b) The charge enclosed by a cylinder of height h, radius R, and base on the xy-plane center at the origin.
We have

$$q_{\text{enc}} = \int_V \rho \, d\tau,$$

with
$$\rho = 2k\varepsilon_o(y + z).$$
We can transform ρ into cylindrical coordinates using $x = s\cos\phi$, $y = s\sin\phi$, and $z = z$.

So
$$q_{enc} = 2k\varepsilon_o \int_0^R \int_0^h \int_0^{2\pi} (s\sin\phi + z)s\,d\phi\,dz\,ds = \pi k\varepsilon_o h^2 R^2.$$

c) The charge enclosed by an upper hemisphere of radius R centered at the origin.

Again
$$q_{enc} = \int_V \rho\,d\tau$$
with
$$\rho = 2k\varepsilon_o(y + z),$$
but now $y = r\sin\phi\sin\theta$ and $z = r\cos\theta$. So
$$q_{enc} = 2k\varepsilon_o \int_0^R \int_0^{2\pi} \int_0^{\frac{\pi}{2}} r(\sin\phi\sin\theta + \cos\theta)r^2\sin\theta\,d\theta\,d\phi\,dr = \frac{k\varepsilon_o\pi R^4}{2}.$$

Problem 2.6. Given a charge q located in the center of a spherical shell of radius R and surface charge $\sigma = k\sin\theta$ (with constant k), find the electric field inside and outside the shell.

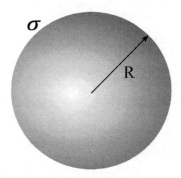

Solution We will use Gauss's law to find the field
$$\oint_S \vec{E} \cdot d\vec{a} = \frac{q_{enc}}{\varepsilon_o},$$

where

$$\oint_S \vec{E} \cdot d\vec{a} = \oint_S E \, da = E \oint_S da = E 4\pi r^2.$$

For $r < R$, we have

$$q_{enc} = q.$$

So

$$E 4\pi r^2 = \frac{q}{\varepsilon_o} \to \vec{E} = \frac{1}{4\pi\varepsilon_o} \frac{q}{r^2} \hat{r}.$$

For $r > R$, we have

$$q_{enc} = q + \int \sigma \, da = q + k \int_0^{2\pi} \int_0^{\pi} R^2 \sin^2\theta \, d\theta \, d\phi.$$

$$q_{enc} = q + kR^2\pi^2.$$

So

$$E 4\pi r^2 = \frac{q}{\varepsilon_o} + \frac{kR^2\pi^2}{\varepsilon_o}$$

and

$$\vec{E} = \frac{1}{4\pi\varepsilon_o} \frac{q + kR^2\pi^2}{r^2} \hat{r}.$$

Problem 2.7. Given a line of charge carrying λ surrounded by a cylindrical shell with inner radius a, outer radius b, and charge density $\rho = ks^2$, find the electric field in the regions $s < a$, $a < s < b$, and $b < s$.

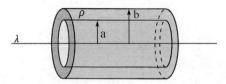

Solution Gauss's law states

$$\oint_S \vec{E} \cdot d\vec{a} = \frac{q_{enc}}{\varepsilon_o},$$

where
$$\oint_S \vec{E} \cdot d\vec{a} = \oint_S E \, da = E \oint_S da = E 2\pi s \ell.$$

For $s < a$, we have
$$q_{\text{enc}} = \lambda \ell.$$

So
$$E 2\pi s \ell = \frac{\lambda \ell}{\varepsilon_o}$$

and
$$\vec{E} = \frac{\lambda}{2\pi \varepsilon_o s} \hat{s}.$$

For $a < s < b$, we have
$$q_{\text{enc}} = \lambda \ell + \int \rho \, d\tau = \lambda \ell + \int_0^\ell \int_0^{2\pi} \int_a^s k(s')^2 s' \, ds' \, d\phi' \, dz' = \ell \left[\lambda + \frac{k\pi}{2}(s^4 - a^4) \right].$$

So
$$\oint_S \vec{E} \cdot d\vec{a} = \frac{q_{\text{enc}}}{\varepsilon_o} \rightarrow E 2\pi s \ell = \frac{\ell \left[\lambda + \frac{k\pi}{2}(s^4 - a^4) \right]}{\varepsilon_o}$$

and
$$\vec{E} = \frac{2\lambda + k\pi(s^4 - a^4)}{4\pi \varepsilon_o s} \hat{s}.$$

For $b < s$, we have
$$q_{\text{enc}} = \lambda \ell + \int \rho \, d\tau = \lambda \ell + \int_0^\ell \int_0^{2\pi} \int_a^b k(s')^2 s' \, ds' \, d\phi' \, dz' = \ell \left[\lambda + \frac{k\pi}{2}(b^4 - a^4) \right].$$

So
$$\oint_S \vec{E} \cdot d\vec{a} = \frac{q_{\text{enc}}}{\varepsilon_o} \rightarrow E 2\pi s \ell = \frac{\ell \left[\lambda + \frac{k\pi}{2}(b^4 - a^4) \right]}{\varepsilon_o}$$

and

$$\vec{E} = \frac{2\lambda + k\pi(b^4 - a^4)}{4\pi\varepsilon_o s}\hat{s}.$$

Problem 2.8. Which of the following is a possible electrostatic field?
a) $\vec{E} = k(yz\hat{x} + xz\hat{y} + x^2\hat{z})$
b) $\vec{E} = k(x\hat{x} + y\hat{y} + z\hat{z})$
c) $\vec{E} = k[2xz\hat{x} + z^2\hat{y} + (x^2 + 2yz)\hat{z}]$

where k is a constant with the appropriate units for the given field. For the possible electric field, find the electric potential using the origin as your reference point. Check your answer by verifying that $\vec{E} = -\nabla V$.

Solutions

a) $\vec{E} = k\left(yz\hat{x} + xz\hat{y} + x^2\hat{z}\right)$

$$\nabla \times \vec{E} = \begin{vmatrix} \hat{x} & \hat{y} & \hat{z} \\ \dfrac{\partial}{\partial x} & \dfrac{\partial}{\partial y} & \dfrac{\partial}{\partial z} \\ yz & xz & x^2 \end{vmatrix}$$

$$= \left[\frac{\partial}{\partial y}(x^2) - \frac{\partial}{\partial z}(xz)\right]\hat{x} + \left[\frac{\partial}{\partial z}(yz) - \frac{\partial}{\partial x}(x^2)\right]\hat{y} + \left[\frac{\partial}{\partial x}(xz) - \frac{\partial}{\partial y}(yz)\right]\hat{z}$$

$$= (0 - x)\hat{x} + (y - 2x)\hat{y} + (z - z)\hat{z} = -x\hat{x} + (-2x + y)\hat{y}.$$

Since $\nabla \times \vec{E} \neq 0$, this is *not* a possible electric field.

b) $\vec{E} = k(x\hat{x} + y\hat{y} + z\hat{z})$

$$\nabla \times \vec{E} = \begin{vmatrix} \hat{x} & \hat{y} & \hat{z} \\ \dfrac{\partial}{\partial x} & \dfrac{\partial}{\partial y} & \dfrac{\partial}{\partial z} \\ x & y & z \end{vmatrix}$$

$$= \left[\frac{\partial}{\partial y}(z) - \frac{\partial}{\partial z}(y)\right]\hat{x} + \left[\frac{\partial}{\partial z}(x) - \frac{\partial}{\partial x}(z)\right]\hat{y} + \left[\frac{\partial}{\partial x}(y) - \frac{\partial}{\partial y}(x)\right]\hat{z} = 0.$$

Since $\nabla \times \vec{E} = 0$, this is a possible electric field. Let us find the electric potential by integrating along the path given by

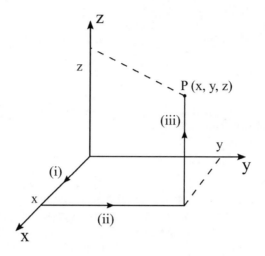

The potential is given by

$$V = -\int_{\mathcal{O}}^{\vec{r}} \vec{E} \cdot d\vec{\ell}.$$

where

$$\vec{E} \cdot d\vec{\ell} = k(x\hat{x} + y\hat{y} + z\hat{z}) \cdot (dx\hat{x} + dy\hat{y} + dz\hat{z})$$
$$= k(x\,dx + y\,dy + z\,dz).$$

Note along (i) we only have dx, along (ii) we only have dy, and along (iii) we only have dz. Therefore, taking the origin $\mathcal{O}=(0, 0, 0)$ as our reference point, the potential will be given by

$$V = -\int_{\mathcal{O}}^{\vec{r}} \vec{E} \cdot d\vec{\ell} = -k \left(\underbrace{\int_0^x x'dx'}_{(i)} + \underbrace{\int_0^y y'dy'}_{(ii)} + \underbrace{\int_0^z z'dz'}_{(iii)} \right).$$

So,

$$V(r) = -\frac{k}{2}\left(x^2 + y^2 + z^2\right).$$

We can check this using
$$\vec{E} = -\nabla V$$
$$= -\left\{\frac{\partial}{\partial x}\left[-\frac{k}{2}(x^2+y^2+z^2)\right]\hat{x} + \frac{\partial}{\partial y}\left[-\frac{k}{2}(x^2+y^2+z^2)\right]\hat{y}\right.$$
$$\left. + \frac{\partial}{\partial z}\left[-\frac{k}{2}(x^2+y^2+z^2)\right]\hat{z}\right\}$$
$$= \frac{2kx}{2}\hat{x} + \frac{2ky}{2}\hat{y} + \frac{2kz}{2}\hat{z}$$
$$= k(x\hat{x} + y\hat{y} + z\hat{z}) = \vec{E}.$$

c) $\vec{E} = k[2xz\hat{x} + z^2\hat{y} + (x^2+2yz)\hat{z}]$

$$\nabla \times \vec{E} = \begin{vmatrix} \hat{x} & \hat{y} & \hat{z} \\ \dfrac{\partial}{\partial x} & \dfrac{\partial}{\partial y} & \dfrac{\partial}{\partial z} \\ 2xz & z^2 & x^2+2yz \end{vmatrix}$$

$$= \left[\frac{\partial}{\partial y}(x^2+2yz) - \frac{\partial}{\partial z}(z^2)\right]\hat{x} + \left[\frac{\partial}{\partial z}(2xz) - \frac{\partial}{\partial x}(x^2+2yz)\right]\hat{y}$$
$$+ \left[\frac{\partial}{\partial x}(z^2) - \frac{\partial}{\partial y}(2xz)\right]\hat{z}$$
$$= k\big[(2z - 2z)\hat{x} + (2x - 2x)\hat{y} + (0 - 0)\hat{z}\big] = 0.$$

Since $\nabla \times \vec{E} = 0$, this is a possible electric field. Let us find the electric potential by integrating along the same path as before. The potential is given by
$$V = -\int_{\mathcal{O}}^{\vec{r}} \vec{E} \cdot d\vec{\ell}$$
where
$$\vec{E} \cdot d\vec{\ell} = k\left[2xz\hat{x} + z^2\hat{y} + (x^2+2yz)\hat{z}\right] \cdot (dx\hat{x} + dy\hat{y} + dz\hat{z})$$
$$= k\left[2xz\,dx + z^2\,dy + (x^2+2yz)dz\right].$$

Note along (i) we only have dx with $y = 0$ and $z = 0$, along (ii) we only have dy with $x = 1$ and $z = 0$, and along (iii) we only have dz with $x = 1$ and $y = 1$. Therefore, taking the origin $\mathcal{O} = (0, 0, 0)$ as our reference point, the potential will be given by

$$V = -\int_{\mathcal{O}}^{\vec{r}} \vec{E} \cdot d\vec{\ell} = -k\left(\underbrace{\int_0^x 2x'z\,dx'}_{(i)} + \underbrace{\int_0^y z^2 dy'}_{(ii)} + \underbrace{\int_0^z (2yz' + x^2)dz'}_{(iii)}\right)$$

$$= -k\left(\underbrace{\int_0^x 2x'(0)dx'}_{(i)} + \underbrace{\int_0^y 0^2\,dy'}_{(ii)} + \underbrace{\int_0^z (2(1)z'+1^2)dz'}_{(iii)}\right)$$

$$V(r) = -k\left(yz^2 + x^2 z\right).$$

We can check this using

$$\vec{E} = -\nabla V$$
$$= -\left\{\frac{\partial}{\partial x}\left[-k\left(yz^2 + x^2 z\right)\right]\hat{x} + \frac{\partial}{\partial y}\left[-k\left(yz^2 + x^2 z\right)\right]\hat{y} + \frac{\partial}{\partial z}\left[-k\left(yz^2 + x^2 z\right)\right]\hat{z}\right\}$$
$$= k\left[(2xz)\hat{x} + \left(z^2\right)\hat{y} + \left(x^2 + 2yz\right)\hat{z}\right] = \vec{E}.$$

Problem 2.9. Find the electric field and the electric potential inside and outside a thin spherical shell of radius R that carries a uniform surface charge σ. Set the reference point at infinity.

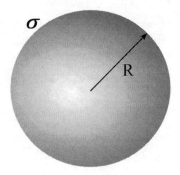

Solution Let us find the electric field everywhere by using Gauss's law, given by

$$\oint_S \vec{E} \cdot d\vec{a} = \frac{q_{\text{enc}}}{\varepsilon_0},$$

where
$$\oint_S \vec{E} \cdot d\vec{a} = \oint_S E \, da = E \oint_S da = E 4\pi r^2.$$

For $r < R$, we have our Gaussian surface given by

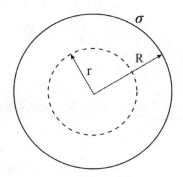

where r is the radius of the Gaussian sphere with radius smaller than R. Note that
$$q_{\text{enc}} = 0.$$
So,
$$E 4\pi r^2 = 0 \rightarrow \vec{E} = 0.$$

For $r > R$, we have our Gaussian surface given by

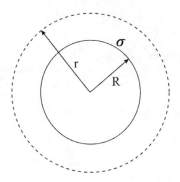

Now we have
$$q_{\text{enc}} = \sigma 4\pi R^2.$$
So,
$$E 4\pi r^2 = \frac{\sigma 4\pi R^2}{\varepsilon_0}$$

and
$$\vec{E} = \frac{\sigma R^2}{\varepsilon_0 r^2}\hat{r}.$$

Now let us calculate the electric potential everywhere taking the reference point at infinity. We will use
$$V = -\int_\infty^{\vec{r}} \vec{E} \cdot d\vec{\ell},$$
where
$$d\vec{\ell} = dr\,\hat{r} + r\,d\theta\,\hat{\theta} + r\sin\theta\,d\phi\,\hat{\phi}.$$

For $r > R$,

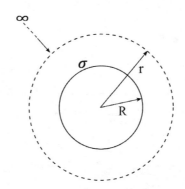

therefore,
$$V = -\int_\infty^{\vec{r}} \vec{E} \cdot d\vec{\ell} = -\int_\infty^{\vec{r}} \frac{\sigma R^2}{\varepsilon_0 r^2}\hat{r} \cdot \left(dr\,\hat{r} + r\,d\theta\,\hat{\theta} + r\sin\theta\,d\phi\,\hat{\phi}\right) = -\int_\infty^{r} \frac{\sigma R^2}{\varepsilon_0 r'^2}dr'$$

$$V = \frac{\sigma R^2}{\varepsilon_0 r}.$$

For $r < R$

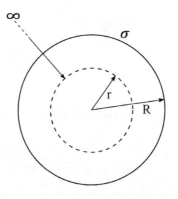

$$V = -\int_\infty^{\vec{r}} \vec{E} \cdot d\vec{\ell} = -\int_\infty^R \frac{\sigma R^2}{\varepsilon_0 r} dr - \int_R^r 0\, dr' = \frac{\sigma R^2}{\varepsilon_0 R} - 0 = \frac{\sigma R}{\varepsilon_0} = \text{const.}$$

Note that the potential inside the shell is constant, as the electric field is zero.

Problem 2.10. Calculate the electric field and the electric potential inside and outside a solid sphere of radius R having a uniform charge distribution ρ. Use infinity as your reference point. Then obtain the gradient of the potential everywhere and check that $\vec{E} = -\nabla V$. Plot the potential versus distance from the center of the sphere.

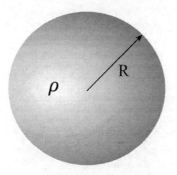

Solution Starting with the electric field, we use Gauss's law, given by

$$\oint_S \vec{E} \cdot d\vec{a} = \frac{q_{\text{enc}}}{\varepsilon_0},$$

where

$$\oint_S \vec{E} \cdot d\vec{a} = \oint_S E\, da = E \oint_S da = E 4\pi r^2.$$

For $r > R$, we have our Gaussian surface given by

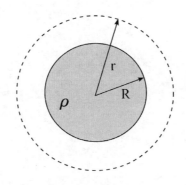

Here we simply have

$$q_{enc} = \rho \mathcal{V}_{sp} = \frac{\rho 4\pi R^3}{3},$$

where $\mathcal{V}_{sp}$ is the volume of the sphere. So

$$E 4\pi r^2 = \frac{\rho 4\pi R^3}{3\varepsilon_0}$$

and

$$\vec{E} = \frac{\rho R^3}{3\varepsilon_0 r^2}\hat{r}.$$

For $r < R$, our Gaussian surface becomes

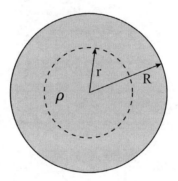

Now,

$$q_{enc} = \rho \mathcal{V}_{enc} = \frac{\rho 4\pi r^3}{3},$$

where $\mathcal{V}_{enc}$ is the enclosed volume. So

$$E 4\pi r^2 = \frac{\rho 4\pi r^3}{3\varepsilon_0}$$

and

$$\vec{E} = \frac{\rho r}{3\varepsilon_0}\hat{r}.$$

The plot of electric field is given by

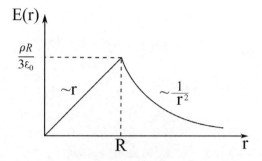

Now we can calculate the electric potential. This is done using

$$V = -\int_{\infty}^{\vec{r}} \vec{E} \cdot d\vec{\ell}$$

with infinity as the reference point. For $r > R$

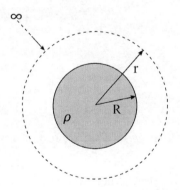

with

$$V = -\int_{\infty}^{\vec{r}} \vec{E} \cdot d\vec{\ell} = -\int_{\infty}^{r} \frac{\rho R^3}{3\varepsilon_0 r'^2} dr' = \frac{\rho R^3}{3\varepsilon_0 r}.$$

For $r < R$

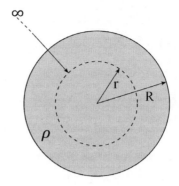

with

$$V = -\int_{\infty}^{\vec{r}} \vec{E} \cdot d\vec{\ell} = -\int_{\infty}^{R} \frac{\rho R^2}{3\varepsilon_0 r^2} dr - \int_{R}^{r} \frac{\rho r'}{3\varepsilon_0} dr' = \frac{\rho R^2}{2\varepsilon_0} - \frac{\rho r^2}{6\varepsilon_0}.$$

We can check using $\vec{E} = -\nabla V$. For $r > R$,

$$\vec{E} = -\nabla V = -\frac{\partial}{\partial r}\left(\frac{\rho R^2}{2\varepsilon_0} - \frac{\rho r^2}{6\varepsilon_0}\right)\hat{r} = \frac{\rho r}{3\varepsilon_0}\hat{r}$$

and for $r < R$,

$$\vec{E} = -\nabla V = -\frac{\partial}{\partial r}\left(\frac{\rho R^3}{2\varepsilon_0 r}\right)\hat{r} = \frac{\rho R^3}{3\varepsilon_0 r^2}\hat{r}$$

both of which are in agreement with what we found from Gauss's law.

Problem 2.11. Calculate the electric field and the electric potential for a sphere of radius R that carries a charge density $\rho = kr^2$, where k is a constant.

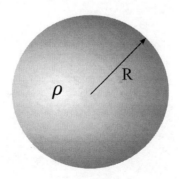

Solution Starting with the electric field, we use Gauss's law, given by

$$\oint_S \vec{E} \cdot d\vec{a} = \frac{q_{enc}}{\varepsilon_0},$$

where

$$\oint_S \vec{E} \cdot d\vec{a} = \oint_S E \, da = E \oint_S da = E 4\pi r^2.$$

For $r > R$, we have our Gaussian surface given by

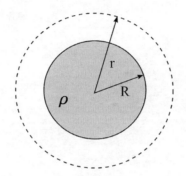

and our enclosed charge is given by

$$q_{enc} = \int \rho \, d\tau = \int_0^{2\pi} d\phi \int_0^{\pi} \sin\theta \, d\theta \int_0^R kr^2 r^2 dr = 4\pi k \frac{R^5}{5}.$$

Therefore,

$$E 4\pi r^2 = \frac{4\pi k R^5}{5\varepsilon_0}$$

so

$$\vec{E} = \frac{kR^5}{5\varepsilon_0 r^2} \hat{r}.$$

For $r < R$, our Gaussian surface becomes

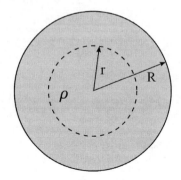

Now,
$$q_{\text{enc}} = \int \rho \, d\tau = \int_0^{2\pi} d\phi \int_0^{\pi} \sin\theta \, d\theta \int_0^R kr'^2 r'^2 dr' = \frac{4\pi k r^5}{5}$$

so
$$E 4\pi r^2 = \frac{4\pi k r^5}{5\varepsilon_0}$$

and
$$\vec{E} = \frac{kr^3}{5\varepsilon_0}\hat{r}.$$

Now we can calculate the electric potential. This is done using
$$V = -\int_\infty^{\vec{r}} \vec{E} \cdot d\vec{\ell}$$

with infinity as the reference point. For $r > R$

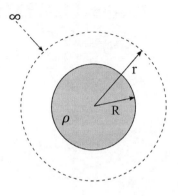

with

$$V = -\int_\infty^{\vec{r}} \vec{E} \cdot d\vec{\ell} = -\int_\infty^r \frac{kR^5}{5\varepsilon_0 r'^2} dr' = \frac{kR^5}{5\varepsilon_0 r}.$$

For $r < R$

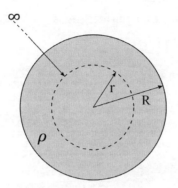

with

$$V = -\int_\infty^{\vec{r}} \vec{E} \cdot d\vec{\ell} = -\int_\infty^R \frac{kR^5}{5\varepsilon_0 r'^2} dr' - \int_R^r \frac{kr'^3}{5\varepsilon_0} dr' = \frac{kR^5}{5\varepsilon_0 R} - 0 - \frac{kr^4}{20\varepsilon_0} + \frac{kR^4}{20\varepsilon_0}$$

$$= \frac{k}{20\varepsilon_0}\left(4R^4 - r^4 + R^4\right)$$

$$V = \frac{kR^4}{20\varepsilon_0}\left(5 - \frac{r^4}{R^4}\right).$$

Problem 2.12 A long cylinder of radius a carries a charge density $\rho = ks^2$, where k is a constant and s is the distance from the axis of the cylinder. Find the electric field and the electric potential everywhere. Take the reference point at a distance b from the axis ($b > a$).

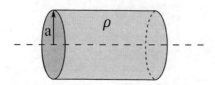

Solution Starting with the electric field, we use Gauss's law, given by

$$\oint_S \vec{E} \cdot d\vec{a} = \frac{q_{enc}}{\varepsilon_0}.$$

Note the left-hand side is always given by

$$\oint_S \vec{E} \cdot d\vec{a} = \oint_S E\, da = E \oint_S da = E 2\pi s \ell.$$

For $s > a$, we have our Gaussian surface given by

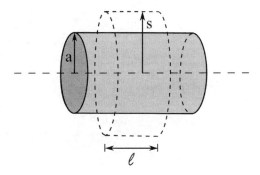

with enclosed charge given by

$$q_{enc} = \int \rho\, d\tau = \int_0^{2\pi} d\phi \int_0^{\ell} dz \int_0^a ks^2 s\, ds = \frac{\pi k \ell a^4}{2}.$$

Therefore,

$$E 2\pi s \ell = \frac{\pi k \ell a^4}{2\varepsilon_0}$$

and

$$\vec{E} = \frac{k a^4}{4\varepsilon_0 s} \hat{s}.$$

For $s < a$, our Gaussian surface becomes

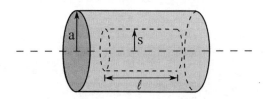

Now,
$$q_{enc} = \int \rho \, d\tau = \int_0^{2\pi} d\phi \int_0^{\ell} dz \int_0^{s} ks'^2 s' \, ds' = \frac{\pi k \ell s^4}{2}$$

so
$$E 2\pi s \ell = \frac{\pi k \ell s^4}{2\varepsilon_0}$$

and
$$\vec{E} = \frac{ks^3}{4\varepsilon_0} \hat{s}.$$

Now we can calculate the electric potential. This is done using
$$V = -\int_{\vec{b}}^{\vec{r}} \vec{E} \cdot d\vec{\ell}$$

with b as the reference point. For $s > a$,

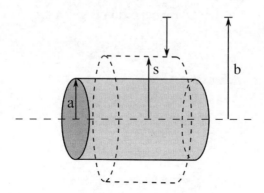

with
$$V = -\int_{\vec{b}}^{\vec{r}} \vec{E} \cdot d\vec{\ell} = -\int_b^s \frac{ka^4}{4\varepsilon_0 s'} ds' = -\frac{ka^4}{4\varepsilon_0}(\ln s - \ln b) = -\frac{ka^4}{4\varepsilon_0} \ln \frac{s}{b}.$$

For $s < a$,

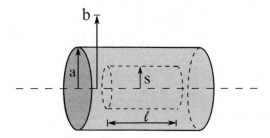

with

$$V = -\int \vec{E} \cdot d\vec{\ell} = -\int_b^a \frac{ka^4}{4\varepsilon_0 s}ds - \int_a^s \frac{ks'^3}{4\varepsilon_0}ds' = -\frac{ka^4}{4\varepsilon_0}\ln\frac{a}{b} - \frac{k(s^4 - a^4)}{16\varepsilon_0}$$

$$V = \frac{ka^4}{4\varepsilon_0}\ln\frac{b}{a} + \frac{k(a^4 - s^4)}{16\varepsilon_0}.$$

Problem 2.13. Verify the electrostatic boundary condition using the charge distribution in problem 2.9.

Solution The electrostatic boundary condition is given by

$$\vec{E}_{above} - \vec{E}_{below} = \frac{\sigma}{\varepsilon_0}\hat{n}.$$

From problem 2.9, our electric fields are

$$\vec{E} = \begin{cases} 0 & r < R \\ \frac{\sigma R^2}{\varepsilon_0 r^2}\hat{r} & r > R \end{cases}.$$

At $r = R$, we have

$$\vec{E}_{above} = \frac{\sigma}{\varepsilon_0}\hat{r}$$

and

$$\vec{E}_{below} = 0.$$

Therefore,

$$\vec{E}_{above} - \vec{E}_{below} = \frac{\sigma}{\varepsilon_0}\hat{r} - 0 = \frac{\sigma}{\varepsilon_0}\hat{n},$$

where $\hat{n}$ is normal to the sphere which has the direction of $\hat{r}$.

Problem 2.14. Find the work required to assemble the charge distribution below.

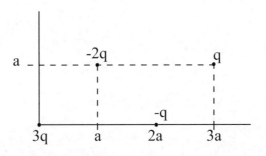

Solution We can denote the following $q_1 = 3q$, $q_2 = -2q$, $q_3 = -q$, $q_4 = q$. Starting with q_1, $W = 0$. Moving in q_2, we have

$$W_2 = \frac{1}{4\pi\varepsilon_0} q_2 \left(\frac{q_1}{r_{12}}\right) = -\frac{6q^2}{4\pi\varepsilon_0} \frac{1}{a\sqrt{2}} = -\frac{q^2}{4\pi\varepsilon_0} \frac{3\sqrt{2}}{a}.$$

Moving in q_3, we have

$$W_3 = \frac{1}{4\pi\varepsilon_0} q_3 \left(\frac{q_1}{r_{13}} + \frac{q_2}{r_{23}}\right) = -\frac{q}{4\pi\varepsilon_0}\left(\frac{3q}{2a} - \frac{2q}{\sqrt{2}a}\right) = \frac{q^2}{4\pi\varepsilon_0}\left(\frac{2\sqrt{2}-3}{2a}\right).$$

Moving in q_4, we have

$$W_4 = \frac{1}{4\pi\varepsilon_0} q_4 \left(\frac{q_1}{r_{14}} + \frac{q_2}{r_{24}} + \frac{q_3}{r_{34}}\right) = \frac{q}{4\pi\varepsilon_0}\left(\frac{3q}{a\sqrt{10}} - \frac{2q}{2a} - \frac{q}{a\sqrt{2}}\right)$$

$$= \frac{q^2}{4\pi\varepsilon_0}\left(\frac{3\sqrt{10} - 5\sqrt{2} - 10}{10a}\right).$$

Therefore,

$$W = W_2 + W_3 + W_4 = \frac{q^2}{4\pi\varepsilon_0}\left(-\frac{3\sqrt{2}}{a} + \frac{2\sqrt{2}-3}{2a} + \frac{3\sqrt{10} - 5\sqrt{2} - 10}{10a}\right)$$

$$W = \frac{q^2}{4\pi\varepsilon_0}\left[\frac{3\sqrt{10} - 25(\sqrt{2}+1)}{10a}\right].$$

Problem 2.15. Find the energy stored in a spherical shell of inner radius a and outer radius b with a charge distribution $\rho = kr^2$.

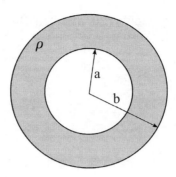

Solution The work is given by

$$W = \frac{\varepsilon_0}{2} \int_{\text{all space}} E^2 \, d\tau$$

so we need to find the field in all three regions.
For $r < a$, we have $q_{\text{enc}} = 0$. So $E = 0$.
For $a < r < b$, we have

$$\oint_S \vec{E} \cdot d\vec{a} = \frac{q_{\text{enc}}}{\varepsilon_0}$$

with

$$q_{\text{enc}} = \int \rho \, d\tau = 4\pi \int_a^r k(r')^2 (r')^2 \, dr' = \frac{4\pi k}{5}(r^5 - a^5)$$

and

$$\oint_S \vec{E} \cdot d\vec{a} = E 4\pi r^2.$$

So

$$E = \frac{k(r^5 - a^5)}{5\varepsilon_0 r^2}.$$

For $r > b$, we have

$$q_{\text{enc}} = \int \rho \, d\tau = 4\pi \int_a^b k(r')^2 (r')^2 \, dr' = \frac{4\pi k}{5}(b^5 - a^5).$$

So
$$E = \frac{k(b^5 - a^5)}{5\varepsilon_0 r^2}.$$

Now the work is given by

$$W = \frac{\varepsilon_0}{2}\left\{4\pi \int_a^b \left[\frac{k(r^5 - a^5)}{5\varepsilon_0 r^2}\right]^2 r^2\, dr + 4\pi \int_b^\infty \left[\frac{k(b^5 - a^5)}{5\varepsilon_0 r^2}\right]^2 r^2\, dr\right\}$$

$$W = \frac{k^2 \pi}{45\varepsilon_0}\left(5a^9 - 9a^5 b^4 + 4b^9\right).$$

Problem 2.16. Given a charge density $\rho = ke^{-r}$, with k a constant, find the radius of a sphere that maximizes the energy per unit volume.

Solution
The energy per unit volume is given by

$$\frac{W}{\text{volume}} = \frac{\varepsilon_0}{2} E^2,$$

where the field is given by

$$\oint_S \vec{E} \cdot d\vec{a} = \frac{q_{enc}}{\varepsilon_0}$$

with

$$q_{enc} = \int \rho\, d\tau = 4\pi \int_0^r ke^{-r'}(r')^2\, dr' = 4\pi k e^{-r}\left(2e^r - r^2 - 2r - 2\right).$$

Also,

$$\oint_S \vec{E} \cdot d\vec{a} = E 4\pi r^2.$$

So

$$E = \frac{ke^{-r}\left(2e^r - r^2 - 2r - 2\right)}{\varepsilon_0 r^2}.$$

The energy per unit volume contained in a sphere of radius r is given by

$$\frac{W}{\text{volume}} = \frac{\varepsilon_0}{2} E^2 = \frac{k^2 e^{-2r}\left(2e^r - r^2 - 2r - 2\right)^2}{2\varepsilon_0 r^4}.$$

To maximize this, we have

$$\frac{d}{dr}\left(\frac{W}{\text{volume}}\right) = 0$$

so

$$\frac{-k^2 e^{-2r}\left(2e^r - r^2 - 2r - 2\right)\left(4e^r - r^3 - 2r^2 - 4r - 4\right)}{\varepsilon_0 r^5} = 0.$$

Since $r \neq 0$, $k \neq 0$, and $e^{-2r} \neq 0$, we have

$$2e^r - r^2 - 2r - 2 = 0 \rightarrow r = 0$$

and

$$4e^r - r^3 - 2r^2 - 4r - 4 = 0 \rightarrow r = 0, \quad r = 1.45123.$$

But $r \neq 0$, so a sphere of radius $r = 1.45123$ has the maximum energy per unit volume.

Problem 2.17. A metal sphere of radius R and charge q is surrounded by two concentric metal shells.
a) Obtain the surface charge density σ at R, a, b, c, and d.
b) Calculate the potential at the center of the sphere by taking infinity as the reference point.

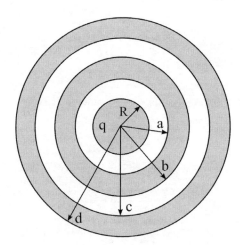

Solutions
a) Obtain the surface charge density σ at R, a, b, c, and d.
For $r = R$, the sphere is metallic, therefore all charge q is distributed on the surface of the sphere. This gives a surface charge density of

$$\sigma = \frac{q}{4\pi R^2}$$

By influence and due to the sphere with charge q, the inner shell is redistributing the electric charge such that the surface with radius a has charge $-q$ and the surface with radius b has charge $+q$. Similarly for the outer shell.
Therefore, the surface charge densities are

$$\sigma_a = -\frac{q}{4\pi a^2} \quad \sigma_b = \frac{q}{4\pi b^2} \quad \sigma_c = -\frac{q}{4\pi c^2} \quad \sigma_d = \frac{q}{4\pi d^2}.$$

b) Calculate the potential at the center of the sphere by taking infinity as the reference point

Taking our reference point at infinity, the electric potential at the center is given by

$$V = -\int_{\infty}^{0} \vec{E} \cdot d\vec{\ell}$$

$$= \int_{\infty}^{d} \frac{q}{4\pi\varepsilon_0 r^2} dr - \int_{d}^{c} 0 \, dr - \int_{c}^{b} \frac{q}{4\pi\varepsilon_0 r^2} dr - \int_{b}^{a} 0 \, dr - \int_{a}^{R} \frac{q}{4\pi\varepsilon_0 r^2} dr - \int_{R}^{0} 0 \, dr$$

$$V = \frac{q}{4\pi\varepsilon_0}\left(\frac{1}{d} + \frac{1}{b} - \frac{1}{c} + \frac{1}{R} - \frac{1}{a}\right).$$

Problem 2.18. Calculate the capacitance of the spherical shell capacitor of radii a (inner) and b (outer) shown below.

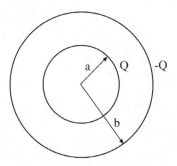

Solution The electric field is due to the inner charge, so

$$E = \frac{Q}{4\pi\varepsilon_0 r^2}.$$

The electric potential is then

$$V = -\int_a^b \vec{E} \cdot d\vec{\ell} = \int_b^a \frac{Q}{4\pi\varepsilon_0 r^2} dr = \frac{Q}{4\pi\varepsilon_0}\left(\frac{1}{a} - \frac{1}{b}\right) = \frac{Q}{4\pi\varepsilon_0}\frac{(b-a)}{ab}.$$

We can find capacitance using

$$V = \frac{Q}{C} \rightarrow C = \frac{Q}{V}.$$

So,

$$C = \frac{Q}{\frac{Q}{4\pi\varepsilon_0}\frac{(b-a)}{ab}} = \frac{4\pi\varepsilon_0 ab}{b-a}.$$

Problem 2.19. Calculate the capacitance of a cylindrical capacitor of length L with two metal cylinders of radii a (inner) and b (outer) shown below. Ignore the edge effects, obtain the capacitance per unit length.

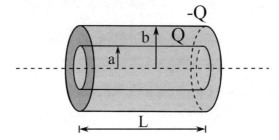

Solution Let us consider that the charge on the inner cylinder is Q (at the radius a). The electric field is obtained from Gauss's law

$$\oint_S \vec{E} \cdot d\vec{a} = \frac{q_{enc}}{\varepsilon_0},$$

where

$$\oint_S \vec{E} \cdot d\vec{a} = \oint_S E \, da = E \oint_S da = E 2\pi s L$$

and the enclosed charge is

$$q_{enc} = Q.$$

Therefore,
$$\vec{E} = \frac{Q}{2\pi\varepsilon_0 s L}\hat{s}.$$

The potential difference between the two cylinders is then
$$V(b) - V(a) = -\int_a^b \vec{E} \cdot d\vec{\ell} = -\int_a^b \frac{Q}{2\pi\varepsilon_0 s L} ds = \frac{Q}{2\pi\varepsilon_0 L} \ln\frac{b}{a}.$$

Therefore, the capacitance is given by
$$C = \frac{Q}{V} = \frac{2\pi\varepsilon_0 L}{\ln\frac{b}{a}}.$$

Bibliography

Byron F W and Fuller R W 1992 *Mathematics of Classical and Quantum Physics* (New York: Dover)
Griffiths D J 1999 *Introduction to Electrodynamics* 3rd edn (Englewood Cliffs, NJ: Prentice Hall)
Griffiths D J 2013 *Introduction to Electrodynamics* 4th edn (New York: Pearson)
Halliday D, Resnick R and Walker J 2010 *Fundamentals of Physics* 9th edn (New York: Wiley)
Halliday D, Resnick R and Walker J 2013 *Fundamentals of Physics* 10th edn (New York: Wiley)
Jackson J D 1998 *Classical Electrodynamics* 3rd edn (New York: Wiley)
Rogawski J 2011 *Calculus: Early Transcendentals* 2nd edn (San Fransisco, CA: Freeman)

IOP Concise Physics

Electromagnetism
Problems and solutions
Carolina C Ilie and Zachariah S Schrecengost

Chapter 3

Electric potential

Chapter 3 contains different methods for obtaining the electric potential. We will focus on calculating the potential as finding the field is a straightforward calculation once the potential has been determined. Laplace's equation is solved using different methods, depending on the type of charge distribution and on the symmetry of the problem. The method of images, separation of variables, and multipole (in particular dipole) expansions are discussed using appropriate examples.

3.1 Theory

3.1.1 Laplace's equation

Cartesian

$$\nabla^2 T = \frac{\partial^2 T}{\partial x^2} + \frac{\partial^2 T}{\partial y^2} + \frac{\partial^2 T}{\partial z^2} = 0$$

Cylindrical

$$\nabla^2 T = \frac{1}{s}\frac{\partial}{\partial s}\left(s\frac{\partial T}{\partial s}\right) + \frac{1}{s^2}\frac{\partial^2 T}{\partial \phi^2} + \frac{\partial^2 T}{\partial z^2} = 0$$

Spherical

$$\nabla^2 T = \frac{1}{r^2}\frac{\partial}{\partial r}\left(r^2\frac{\partial T}{\partial r}\right) + \frac{1}{r^2 \sin\theta}\frac{\partial}{\partial \theta}\left(\sin\theta \frac{\partial T}{\partial \theta}\right) + \frac{1}{r^2 \sin^2\theta}\frac{\partial^2 T}{\partial \phi^2} = 0$$

3.1.2 Solving Laplace's equation

As an introduction to solving problems using Laplace's equation, we will outline the solutions in Cartesian and spherical coordinates. Laplace's equation can be solved by the method separation of variables when we know the boundary conditions. The general solutions will be outlined below, but seeing how they are derived is important. We will leave them in a general form and problems in this chapter will provide examples of using the boundary conditions to solve for the constants.

Two-dimensional Cartesian coordinates
Let us look at a general two-dimensional case where Laplace's equation is given by

$$\nabla^2 V = \frac{\partial^2 V}{\partial x^2} + \frac{\partial^2 V}{\partial y^2} = 0.$$

We look for a solution of the type $V(x, y) = X(x)Y(y)$ and we replace the desired solution in Laplace's equation which becomes

$$Y(y)\frac{\partial^2 X(x)}{\partial x^2} + X(x)\frac{\partial^2 Y(y)}{\partial y^2} = 0$$

and in a simpler form

$$Y\frac{\partial^2 X}{\partial x^2} + X\frac{\partial^2 Y}{\partial y^2} = 0.$$

We want to separate the variables, which can easily be done by dividing the equation by $X(x)Y(y) = V(x, y)$

$$\frac{1}{X}\frac{\partial^2 X}{\partial x^2} + \frac{1}{Y}\frac{\partial^2 Y}{\partial y^2} = 0.$$

Note that the first term depends only on x and the second term depends only on y. This means that each of the two terms must be constant, and the two constants must be equal in magnitude but opposite in sign. So

$$\frac{1}{X}\frac{\partial^2 X}{\partial x^2} = F \qquad \frac{1}{Y}\frac{\partial^2 Y}{\partial y^2} = -F.$$

We choose F positive, and we can rewrite the equations as

$$\frac{1}{X}\frac{d^2 X}{dx^2} = k^2 \qquad \frac{1}{Y}\frac{d^2 Y}{dy^2} = -k^2.$$

Note that the initial partial differential equation was replaced by two ordinary differential equations. Rearranging yields

$$\frac{d^2 X}{dx^2} = k^2 X \qquad \frac{d^2 Y}{dy^2} = -k^2 Y.$$

The two equations have the following solutions

$$X(x) = Ae^{kx} + Be^{-kx}$$

and
$$Y(y) = C\sin(ky) + D\cos(ky).$$
Going back to the electric potential, V becomes
$$V(x, y) = X(x)Y(y) = \left(Ae^{kx} + Be^{-kx}\right)\left[C\sin(ky) + D\cos(ky)\right].$$
The next step is to apply the boundary conditions in order to obtain the constants A, B, C, and D and to (usually) impose some constraints on k.

Two-dimensional spherical coordinates
Here we will assume azimuthal symmetry (no dependence on ϕ) where Laplace's equation is given by
$$\nabla^2 V = \frac{1}{r^2}\frac{\partial}{\partial r}\left(r^2\frac{\partial V}{\partial r}\right) + \frac{1}{r^2\sin\theta}\frac{\partial}{\partial \theta}\left(\sin\theta\frac{\partial V}{\partial \theta}\right) = 0.$$
We look for a solution which has a radial component and an angular component
$$V(r, \theta) = R(r)\Theta(\theta).$$
Note that here R is the function of r, and not merely the radius of the sphere. We plug our solution in the previous equation and we obtain
$$\Theta(\theta)\frac{\partial}{\partial r}\left(r^2\frac{\partial R(r)}{\partial r}\right) + \frac{R(r)}{\sin\theta}\frac{\partial}{\partial \theta}\left(\sin\theta\frac{\partial \Theta(\theta)}{\partial \theta}\right) = 0.$$
We want to use the method of separation of variables so we will divide the previous equation by $V(r, \theta) = R(r)\Theta(\theta)$,
$$\frac{1}{R(r)}\frac{d}{dr}\left(r^2\frac{dR(r)}{dr}\right) + \frac{1}{\Theta(\theta)\sin\theta}\frac{d}{d\theta}\left(\sin\theta\frac{d\Theta(\theta)}{d\theta}\right) = 0.$$
Note that each term is only a function of a single variable so we were able to replace the partial derivates with ordinary derivates. Now we have one term in $R(r)$ and another term in $\Theta(\theta)$, so we have separated the variables. Therefore, each term must be constant. For well know reasons (more apparent in quantum mechanics), we choose the constant as following
$$\frac{1}{R(r)}\frac{d}{dr}\left(r^2\frac{dR(r)}{dr}\right) = l(l+1)$$
$$\frac{1}{\Theta(\theta)\sin\theta}\frac{d}{d\theta}\left(\sin\theta\frac{d\Theta(\theta)}{d\theta}\right) = -l(l+1).$$
Now let us analyze each of the equations and find the solution.

The radial equation

$$\frac{d}{dr}\left(r^2 \frac{dR(r)}{dr}\right) = l(l+1)R(r)$$

has the general solution, with A and B constants

$$R(r) = Ar^l + \frac{B}{r^{l+1}}.$$

The angular equation

$$\frac{d}{d\theta}\left(\sin\theta \frac{d\Theta(\theta)}{d\theta}\right) = -l(l+1)\Theta(\theta)\sin\theta$$

is not at all trivial. The solutions constitute Legendre polynomials with the variable $\cos\theta$. Legendre polynomials are a special class of polynomials. So the solution to the angular equation is

$$\Theta(\theta) = P_l(\cos\theta),$$

where the general form is given by Rodrigues formula

$$P_l(x) = \frac{1}{2^l l!}\left(\frac{d}{dx}\right)^l (x^2 - 1)^l.$$

Therefore, the separable solution of the Laplace equation, considering azimuthal symmetry, is

$$V(r, \theta) = R(r)\Theta(\theta) = \left(Ar^l + \frac{B}{r^{l+1}}\right)P_l(\cos\theta)$$

and the general solution is the linear combination of the separable solutions

$$V(r, \theta) = \sum_{l=0}^{\infty}\left(A_l r^l + \frac{B_l}{r^{l+1}}\right)P_l(\cos\theta).$$

3.1.3 General solutions

Cartesian

$$\frac{\partial^2 V}{\partial x^2} + \frac{\partial^2 V}{\partial y^2} = 0 \rightarrow V(x,y) = (Ae^{kx} + Be^{-kx})\left[C\sin(ky) + D\cos(ky)\right]$$

Spherical

$$\frac{1}{r^2}\frac{\partial}{\partial r}\left(r^2 \frac{\partial V}{\partial r}\right) + \frac{1}{r^2 \sin\theta}\frac{\partial}{\partial \theta}\left(\sin\theta \frac{\partial V}{\partial \theta}\right) = 0$$

$$V(r, \theta) = \sum_{l=0}^{\infty} \left(A_l r^l + \frac{B_l}{r^{l+1}} \right) P_l(\cos \theta),$$

where P_l are Legendre polynomials given by the Rodrigues formula

$$P_l(x) = \frac{1}{2^l l!} \left(\frac{d}{dx} \right)^l (x^2 - 1)^l.$$

Note

$$P_0(x) = 1$$
$$P_1(x) = x$$
$$P_2(x) = \frac{3x^2 - 1}{2}$$
$$P_3(x) = \frac{5x^3 - 3x}{2}$$
$$P_4(x) = \frac{35x^4 - 30x^2 + 3}{8}.$$

Cylindrical

$$\frac{1}{s} \frac{\partial}{\partial s} \left(s \frac{\partial V}{\partial s} \right) + \frac{1}{s^2} \frac{\partial^2 V}{\partial \phi^2} + \frac{\partial^2 V}{\partial z^2} = 0$$

$$V(s, \phi) = a_0 + b_0 \ln(s) + \sum_{k=1}^{\infty} \left\{ s^k \left[a_k \cos(k\phi) + b_k \sin(k\phi) \right] \right.$$
$$\left. + s^{-k} \left[c_k \cos(k\phi) + d_k \sin(k\phi) \right] \right\}.$$

3.1.4 Method of images

The method of images is a very useful technique for calculating the electric field and the electric potential for problems with symmetry. By using the uniqueness theorem we know that the electric field is uniquely determined at any point in space, thus we can replace an apparently difficult problem with another problem in which we use the initial charge(s) and also the 'images' of the charges. The boundary conditions need to be fulfilled. Typically, this involves the condition for a zero electric potential for a grounded conductor and the condition for zero potential very far away from the system of charges. The potential can be determined in the permitted region, which is in general the region of the real charge. The region of the image charges is the 'forbidden' region; the potential cannot be calculated there. The best way to learn this method is by solving problems and checking the examples.

3.1.5 Potential due to a dipole

$$V(\vec{r}) \cong \frac{1}{4\pi\varepsilon_0} \frac{qd\cos\theta}{r^2} = \frac{1}{4\pi\varepsilon_0} \frac{\vec{p}\cdot\hat{r}}{r^2}$$

3.1.6 Multiple expansion

$$V(\vec{r}) = \frac{1}{4\pi\varepsilon_0} \sum_{n=0}^{\infty} \frac{1}{r^{n+1}} \int (r')^n P_n(\cos\theta') \rho(\vec{r}')d\tau'$$

$$V(\vec{r}) = \frac{1}{4\pi\varepsilon_0}\left[\frac{1}{r}\int \rho(\vec{r}')d\tau' + \frac{1}{r^2}\int r'\cos\theta'\, \rho(\vec{r}')d\tau' \right.$$

$$\left. + \frac{1}{r^3}\int (r')^2\left(\frac{3}{2}\cos^2\theta' - \frac{1}{2}\right)\rho(\vec{r}')d\tau' + \cdots\right].$$

3.1.7 Monopole moment

$$Q = \sum_{i=0}^{n} q_i$$

3.2 Problems and solutions

Problem 3.1. Solve the Laplace equation in spherical and cylindrical coordinates for the cases where V is only dependent on one coordinate at a time.

Solution In spherical coordinates

$$\nabla^2 V = \frac{1}{r^2}\frac{\partial}{\partial r}\left(r^2\frac{\partial V}{\partial r}\right) + \frac{1}{r^2\sin\theta}\frac{\partial}{\partial\theta}\left(\sin\theta\frac{\partial V}{\partial\theta}\right) + \frac{1}{r^2\sin^2\theta}\frac{\partial^2 V}{\partial\phi^2} = 0.$$

If V only depends on r

$$\frac{1}{r^2}\frac{d}{dr}\left(r^2\frac{dV}{dr}\right) = 0,$$

which means

$$r^2\frac{dV}{dr} = C.$$

So

$$V = C\int r^{-2}\,dr \rightarrow V(r) = C\left(-r^{-1} + A\right) = k - \frac{C}{r}.$$

If V only depends on θ

$$\frac{1}{r^2 \sin\theta} \frac{d}{d\theta}\left(\sin\theta \frac{dV}{d\theta}\right) = 0,$$

which mean

$$\sin\theta \frac{dV}{d\theta} = C.$$

So

$$V = C\int \frac{1}{\sin\theta} d\theta = C\int \csc\theta\, d\theta = C\Big(\ln|\csc\theta - \cot\theta| + A\Big)$$

$$V(\theta) = k + C\ln|\csc\theta - \cot\theta|.$$

If V only depends on ϕ

$$\frac{1}{r^2 \sin^2\theta} \frac{d^2V}{d\phi^2} = 0,$$

which means

$$\frac{dV}{d\phi} = C.$$

So

$$V(\phi) = k + C\phi.$$

In cylindrical coordinates

$$\frac{1}{s} \frac{\partial}{\partial s}\left(s \frac{\partial V}{\partial s}\right) + \frac{1}{s^2} \frac{\partial^2 V}{\partial \phi^2} + \frac{\partial^2 V}{\partial z^2} = 0.$$

If V only depends on s

$$\frac{1}{s} \frac{d}{ds}\left(s \frac{dV}{ds}\right) = 0,$$

which means

$$s \frac{dV}{ds} = C.$$

So

$$V = C\int s^{-1} ds = C\Big(\ln|s| + A\Big)$$

$$V(s) = k + C\ln|s|.$$

If V only depends on ϕ,
$$\frac{1}{s^2}\frac{d^2V}{d\phi^2} = 0,$$
which means
$$\frac{dV}{d\phi} = C,$$
which is the same as ϕ dependence in spherical coordinates. So
$$V(\phi) = k + C\phi.$$

If V only depends on z
$$\frac{d^2V}{dz^2} = 0,$$
which means
$$\frac{dV}{dz} = C,$$
which is the same form as ϕ dependence. So
$$V(z) = k + Cz.$$

Problem 3.2. In two-dimensional Cartesian coordinates, the general solution to the Laplace equation is
$$V(x, y) = \left(Ae^{kx} + Be^{-kx}\right)\left[C\sin(ky) + D\cos(ky)\right].$$
Verify that this does in fact satisfy the Laplace equation.

Solution Here, the Laplace equation is
$$\frac{\partial^2 V}{\partial x^2} + \frac{\partial^2 V}{\partial y^2} = 0.$$
So we need to compute $\frac{\partial^2 V}{\partial x^2}$ and $\frac{\partial^2 V}{\partial y^2}$. We have
$$\frac{\partial V}{\partial x} = \left(Ake^{kx} - Bke^{-kx}\right)\left[C\sin(ky) + D\cos(ky)\right]$$
and
$$\frac{\partial^2 V}{\partial x^2} = \left(Ak^2e^{kx} + Bk^2e^{-kx}\right)\left[C\sin(ky) + D\cos(ky)\right]$$
$$= k^2\left(Ae^{kx} + Be^{-kx}\right)\left[C\sin(ky) + D\cos(ky)\right].$$

Also,
$$\frac{\partial V}{\partial y} = \left(Ae^{kx} + Be^{-kx}\right)\left[Ck\cos(ky) - Dk\sin(ky)\right]$$

and
$$\frac{\partial^2 V}{\partial y^2} = \left(Ae^{kx} + Be^{-kx}\right)\left[-Ck^2\sin(ky) - Dk^2\cos(ky)\right]$$
$$= -k^2\left(Ae^{kx} + Be^{-kx}\right)\left[C\sin(ky) + D\cos(ky)\right].$$

Putting this all together, we have
$$\frac{\partial^2 V}{\partial x^2} + \frac{\partial^2 V}{\partial y^2} = k^2\left(Ae^{kx} + Be^{-kx}\right)\left[C\sin(ky) + D\cos(ky)\right]$$
$$- k^2\left(Ae^{kx} + Be^{-kx}\right)\left[C\sin(ky) + D\cos(ky)\right]$$

$$\frac{\partial^2 V}{\partial x^2} + \frac{\partial^2 V}{\partial y^2} = 0$$

as expected.

Problem 3.3. A charge q is placed at a distance d from an infinite grounded conducting plane. Using the method of images, find the electric potential. Which is the 'forbidden' region, for which we cannot calculate the potential?

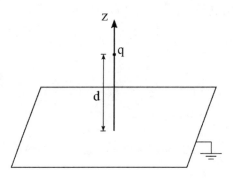

Solution We replace the previous problem with a completely different problem: charge q at $(0, 0, d)$ and its image, charge $-q$ situated at $(0, 0, -d)$. The grounded conducting plane disappeared. This new problem is shown below

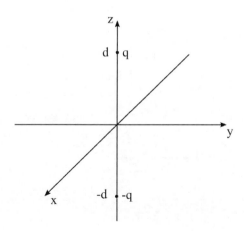

The potential needs to satisfy the following boundary conditions:
a) $V = 0$ for $z = 0$ (grounded plane in the initial problem).
b) $V \to 0$ for a point far from charge q

$$x^2 + y^2 + z^2 \gg d^2.$$

The electric potential due to both point charges is:

$$V(x, y, z) = \frac{q}{4\pi\varepsilon_0\sqrt{x^2 + y^2 + (z-d)^2}} + \frac{-q}{4\pi\varepsilon_0\sqrt{x^2 + y^2 + (z+d)^2}}.$$

Let us check the boundary conditions:
a) For $z = 0$ it is easy to see that $V = 0$.

b) For $x^2 + y^2 + z^2 \gg d^2$, $\sqrt{x^2 + y^2 + (z-d)^2} \cong \sqrt{x^2 + y^2 + (z+d)^2}$, and $V \to 0$.

It is important to note that the only region for which we are able to obtain the electric potential is the region in space above the grounded conducting plane, i.e. the semi-space where charge q is located. For $z < 0$, we are not able to obtain the electric potential.

Problem 3.4. A charge q is placed in an opened grounded conducting parallelepiped at (a, b, c), where a, b, c are positive. Using the method of images, obtain the electric potential in the region of the charge q (for which $x > 0$, $y > 0$, $z > 0$).

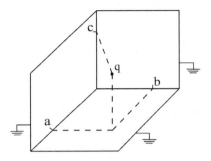

Solution We have the one real charge q at (a, b, c) and seven image charges as following: $-q$ at $(-a, -b, -c)$, $(-a, b, c)$, $(a, -b, c)$, and $(a, b, -c)$; q at $(a, -b, -c)$, $(-a, b, -c)$, $(-a, -b, c)$. The electric potential is given by

$$V(x, y, z)$$
$$= \frac{1}{4\pi\varepsilon_0}\left(\frac{q}{\sqrt{(x-a)^2+(y-b)^2+(z-c)^2}} + \frac{q}{\sqrt{(x-a)^2+(y+b)^2+(z+c)^2}}\right.$$
$$+ \frac{q}{\sqrt{(x+a)^2+(y-b)^2+(z+c)^2}} + \frac{q}{\sqrt{(x+a)^2+(y+b)^2+(z-c)^2}}$$
$$+ \frac{-q}{\sqrt{(x+a)^2+(y+b)^2+(z+c)^2}} + \frac{-q}{\sqrt{(x-a)^2+(y+b)^2+(z-c)^2}}$$
$$+ \frac{-q}{\sqrt{(x-a)^2+(y-b)^2+(z+c)^2}} + \left.\frac{-q}{\sqrt{(x+a)^2+(y-b)^2+(z-c)^2}}\right).$$

If we check the limit $x = 0$, $y = 0$, $z = 0$ successively, we obtain a zero potential for all the three sides of the parallelepiped, where the grounded conductors were in the equivalent problem. Also, for the points far away from the point (a, b, c) in the 'eighth' part of space for which $x \gg 0$, $y \gg 0$, $z \gg 0$, the potential becomes zero as well. Again, note that this is the electric potential only for this part of space, accessible for investigation using the method of images.

Problem 3.5. Let us imagine that we have n charges placed as following: $q_1 = -q$ at $(0, 0, d)$, $q_2 = 2q$ at $(0, 0, 2d)$, ..., $q_n = (-1)^n nq$ at $(0, 0, nd)$ above a grounded, conducting xy-plane (shown below). Obtain the electric potential using the method of images.

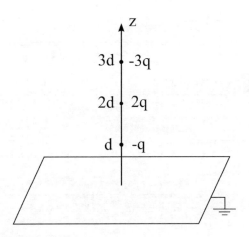

Solution The image charges will be equal in magnitude, of different sign, and situated symmetrically with the xy-plane. In the new problem we eliminate the grounded, conducting plane, but we use the xy-plane for geometrical purposes.

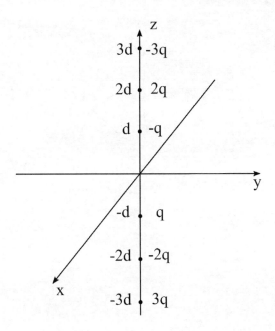

For each charge $q_n = (-1)^n nq$ we have the image charge $q'_n = (-1)^{n+1} nq$ located at $(0, 0, -nd)$. The electric potential is, therefore,

$$V(x, y, z) = \frac{q}{4\pi\varepsilon_0}\left(\frac{-1}{\sqrt{x^2+y^2+(z-d)^2}} + \frac{1}{\sqrt{x^2+y^2+(z+d)^2}} \right.$$
$$+ \frac{(-1)^2 2}{\sqrt{x^2+y^2+(z-2d)^2}} + \frac{(-1)^{2+1} 2}{\sqrt{x^2+y^2+(z+2d)^2}} + \cdots$$
$$+ \frac{(-1)^k k}{\sqrt{x^2+y^2+(z-kd)^2}} + \frac{(-1)^{k+1} k}{\sqrt{x^2+y^2+(z+kd)^2}}$$
$$\left. + \cdots + \frac{(-1)^n n}{\sqrt{x^2+y^2+(z-nd)^2}} + \frac{(-1)^{n+1} n}{\sqrt{x^2+y^2+(z+nd)^2}} \right).$$

It is easy to see that $V = 0$ for $z = 0$ and also $V = 0$ for a point very far from charge $x^2 + y^2 + z^2 \gg (nd)^2$.

Problem 3.6. A conducting sphere of radius R, centered at the origin, is grounded. Find the potential outside the sphere, if a point charge $+q$ is placed at a distance d from the sphere, $d > R$. Use the method of images.

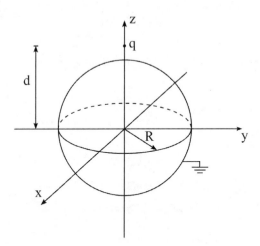

Solution We replace our problem with the grounded, conducting sphere of radius R and the charge $+q$ at distance $d > R$ with a different problem. The sphere, the charge $+q$ and the image charge q', situated at $(0, 0, a)$, with $a < R$, is given by

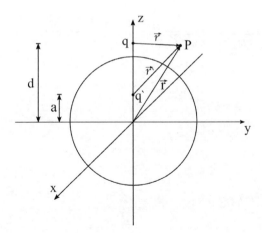

We need $V = 0$ everywhere on the sphere of radius R. Note that we can only find the electric potential outside the sphere. Consider the point P depicted above; here the electric potential at P is given by

$$V(\vec{r}) = \frac{1}{4\pi\varepsilon_0}\left(\frac{q}{r_1} + \frac{q'}{r_2}\right) = \frac{1}{4\pi\varepsilon_0}\left(\frac{q}{|\vec{r}-\vec{d}|} + \frac{q'}{|\vec{r}-\vec{a}|}\right).$$

Using the law of cosines, we have

$$r_1^2 = d^2 + r^2 - 2dr\cos\theta$$

and

$$r_2^2 = a^2 + r^2 - 2ar\cos\theta.$$

We can rewrite V as

$$V(\vec{r}) = \frac{1}{4\pi\varepsilon_0}\left(\frac{q}{\sqrt{d^2 + r^2 - 2dr\cos\theta}} + \frac{q'}{\sqrt{a^2 + r^2 - 2ar\cos\theta}}\right).$$

The potential should be zero for $r = R$

$$V(R) = \frac{1}{4\pi\varepsilon_0}\left(\frac{q}{\sqrt{d^2 + R^2 - 2dR\cos\theta}} + \frac{q'}{\sqrt{a^2 + R^2 - 2aR\cos\theta}}\right) = 0.$$

So

$$\frac{q}{\sqrt{d^2 + R^2 - 2Rd\cos\theta}} = \frac{-q'}{\sqrt{a^2 + R^2 - 2aR\cos\theta}}.$$

We need to obtain both q' and a, so we need two equations. We choose two convenient values for θ, $\theta = 0$ and $\theta = \pi$. For $\theta = 0$, $\cos\theta = 1$, so

$$\frac{q}{\sqrt{d^2 + R^2 - 2Rd}} = \frac{-q'}{\sqrt{a^2 + R^2 - 2aR}} \rightarrow \frac{q}{\sqrt{(d-R)^2}} = \frac{-q'}{\sqrt{(R-a)^2}}.$$

Choosing the positive square root,

$$\frac{q}{d - R} = \frac{-q'}{R - a}$$

and solving for the image charge, we obtain

$$q' = -q \frac{R - a}{d - R}.$$

For $\theta = \pi$, $\cos\theta = -1$, so

$$\frac{q}{d + R} = \frac{-q'}{R + a}.$$

By substituting q', we have

$$\frac{q}{d + R} = \frac{(R - a)q}{(d - R)(a + R)}.$$

Solving for a, we obtain

$$a = \frac{R^2}{d}.$$

Using this, we can find our image charge

$$q' = -\frac{\left(R - \frac{R^2}{d}\right)q}{d - R} = -\frac{R(d - R)q}{(d - R)d}$$

$$q' = -\frac{R}{d}q.$$

Now we have the electric potential, since we obtained the image charge and its position.

Problem 3.7. Given two infinitely long grounded plates at $y = 0$ and $y = a$ connected by the metal strip at $x = -b$ with constant potential $-V_0$ and $x = b$ with constant potential V_0. Find the potential inside the pipe.

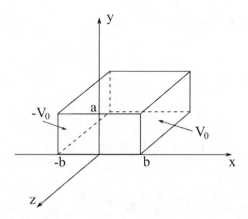

Solution This is independent of z so we have

$$\frac{\partial^2 V}{\partial x^2} + \frac{\partial^2 V}{\partial y^2} = 0$$

with boundary conditions
 (i) $V(y = 0) = 0$
 (ii) $V(y = a) = 0$
 (iii) $V(x = b) = V_0$
 (iv) $V(x = -b) = -V_0$.

Our general solution is given by

$$V(x, y) = \left(Ae^{kx} + Be^{-kx}\right)\left[C\sin(ky) + D\cos(ky)\right].$$

From boundary condition (i)

$$V(x, 0) = \left(Ae^{kx} + Be^{-kx}\right)(D) = 0 \Rightarrow D = 0.$$

So our solution becomes

$$V(x, y) = \left(Ae^{kx} + Be^{-kx}\right)\left[C\sin(ky)\right].$$

From boundary condition (ii)

$$V(x, y) = \left(Ae^{kx} + Be^{-kx}\right)\left[C\sin(ka)\right] = 0$$

we have

$$k = \frac{n\pi}{a},$$

where n is an integer. By symmetry
$$V(x, y) = -V(-x, y)$$
$$\left(Ae^{kx} + Be^{-kx}\right)[C\sin(ka)] = -\left(Ae^{-kx} + Be^{kx}\right)[C\sin(ka)]$$
$$Ae^{kx} + Be^{-kx} = -Ae^{-kx} + Be^{kx}$$
$$A\left(e^{kx} + e^{-kx}\right) = -B\left(e^{kx} + e^{-kx}\right).$$

So
$$A = -B.$$

Absorbing our constants, the solution becomes
$$V(x, y) = C\left(e^{kx} + e^{-kx}\right)\sin(ky) = C\sinh(kx)\sin(ky)$$
and in general
$$V(x, y) = \sum_{n=1}^{\infty} C_n \sinh\left(\frac{n\pi x}{a}\right)\sin\left(\frac{n\pi y}{a}\right).$$

Now to find C_n we take
$$V(b, y) = \sum_{n=1}^{\infty} C_n \sinh\left(\frac{n\pi b}{a}\right)\sin\left(\frac{n\pi y}{a}\right) = V_0$$

so
$$\sum_{n=1}^{\infty} C_n \sinh\left(\frac{n\pi b}{a}\right)\int_0^a \sin\left(\frac{n\pi y}{a}\right)\sin\left(\frac{n'\pi y}{a}\right)dy = \int_0^a V_0 \sin\left(\frac{n'\pi y}{a}\right)dy.$$

Note that when $n \neq n'$,
$$\int_0^a \sin\left(\frac{n\pi y}{a}\right)\sin\left(\frac{n'\pi y}{a}\right)dy = 0$$
and when $n = n'$,
$$\int_0^a \sin\left(\frac{n\pi y}{a}\right)\sin\left(\frac{n'\pi y}{a}\right)dy = \frac{a}{2}.$$

Therefore,
$$C_n \sinh\left(\frac{n\pi b}{a}\right)\frac{a}{2} = \frac{V_0 a}{n\pi}[1 - \cos(n\pi)] = \begin{cases} 0 & n \text{ is even} \\ \frac{2V_0 a}{n\pi} & n \text{ is odd} \end{cases}.$$

3-17

So
$$C_n = \frac{4V_0}{n\pi} \frac{1}{\sinh\left(\frac{n\pi b}{a}\right)}$$

for $n = 1, 3, 5, \ldots$ and our potential is given by

$$V(x, y) = \frac{4V_0}{\pi} \sum_{n=1,3,5}^{\infty} \frac{1}{n} \frac{\sinh\left(\frac{n\pi x}{a}\right)}{\sinh\left(\frac{n\pi b}{a}\right)} \sin\left(\frac{n\pi y}{a}\right).$$

Problem 3.8. Suppose a thin spherical shell of radius R has potential $V(\theta) = V_0(1 - \frac{3}{2}\sin^2\theta)$ specified at the surface. Find the potential inside and outside the sphere.

Solution Our general solution is given by
$$V(r, \theta) = \sum_{l=0}^{\infty}\left(A_l r^l + \frac{B_l}{r^{l+1}}\right)P_l(\cos\theta).$$

Inside:
Here, we must have $B_l = 0$ so the potential does not blow up at the origin. So our potential becomes
$$V(r, \theta) = \sum_{l=0}^{\infty} A_l r^l P_l(\cos\theta).$$

At the surface, we have
$$V(R, \theta) = \sum_{l=0}^{\infty} A_l R^l P_l(\cos\theta) = V_0\left(1 - \frac{3}{2}\sin^2\theta\right).$$

Note that
$$V_0\left(1 - \frac{3}{2}\sin^2\theta\right) = V_0\left[1 - \frac{3}{2}(1 - \cos^2\theta)\right] = V_0\left(\frac{3\cos^2\theta - 1}{2}\right) = V_0 P_2(\cos\theta).$$

So
$$\sum_{l=0}^{\infty} A_l R^l P_l(\cos\theta) = V_0 P_2(\cos\theta).$$

This means we only have the $l = 2$ term,
$$A_2 R^2 P_2(\cos\theta) = V_0 P_2(\cos\theta) \rightarrow A_2 = \frac{V_0}{R^2}.$$

Therefore,
$$V(r, \theta) = A_2 r^2 P_2(\cos\theta) = V_0\left(\frac{r}{R}\right)^2\left(\frac{3\cos^2\theta - 1}{2}\right).$$

Outside:
Here, we must have $A_l = 0$ so the potential does not blow up as $r \to \infty$. So our potential becomes

$$V(r, \theta) = \sum_{l=0}^{\infty} \frac{B_l}{r^{l+1}} P_l(\cos \theta).$$

At the surface, we have

$$V(R, \theta) = \sum_{l=0}^{\infty} \frac{B_l}{R^{l+1}} P_l(\cos \theta) = V_0\left(1 - \frac{3}{2}\sin^2 \theta\right).$$

Again, the right-hand side is $V_0 P_2(\cos \theta)$, so

$$\frac{B_2}{R^{2+1}} P_2(\cos \theta) = V_0 P_2(\cos \theta) \to B_2 = V_0 R^3.$$

Therefore,

$$V(r, \theta) = \frac{B_2}{r^3} P_2(\cos \theta) = V_0 \left(\frac{R}{r}\right)^3 \left(\frac{3\cos^2 \theta - 1}{2}\right).$$

Problem 3.9. A spherical shell of radius R has surface charge $\sigma_0(\theta) = \sin \theta \sin 3\theta$ smeared on its surface. Find the potential inside and outside the sphere.

Solution Our general solution is given by

$$V(r, \theta) = \sum_{l=0}^{\infty} \left(A_l r^l + \frac{B_l}{r^{l+1}}\right) P_l(\cos \theta).$$

Inside we must have $B_l = 0$, otherwise $V \to \infty$ as $r \to 0$. So our potential becomes

$$V_{\text{in}}(r, \theta) = \sum_{l=0}^{\infty} A_l r^l P_l(\cos \theta).$$

Outside we must have $A_l = 0$, otherwise $V \to \infty$ as $r \to \infty$. So our potential becomes

$$V_{\text{out}}(r, \theta) = \sum_{l=0}^{\infty} \frac{B_l}{r^{l+1}} P_l(\cos \theta).$$

At the surface they must be equal, so

$$V_{\text{in}}(R, \theta) = V_{\text{out}}(R, \theta)$$

$$\sum_{l=0}^{\infty} A_l R^l P_l(\cos \theta) = \sum_{l=0}^{\infty} \frac{B_l}{R^{l+1}} P_l(\cos \theta)$$

$$B_l = A_l R^{2l+1}.$$

We must also have

$$\left(\frac{\partial V_{\text{out}}}{\partial r} - \frac{\partial V_{\text{in}}}{\partial r}\right)\bigg|_{r=R} = -\frac{1}{\epsilon_0}\sigma_0(\theta),$$

where

$$\frac{\partial V_{\text{out}}}{\partial r} = \sum_{l=0}^{\infty} -(l+1)\frac{B_l}{r^{l+2}}P_l(\cos\theta)$$

and

$$\frac{\partial V_{\text{in}}}{\partial r} = \sum_{l=0}^{\infty} lA_l r^{l-1}P_l(\cos\theta).$$

Thus,

$$\left(\frac{\partial V_{\text{out}}}{\partial r} - \frac{\partial V_{\text{in}}}{\partial r}\right)\bigg|_{r=R} = \sum_{l=0}^{\infty} -(l+1)\frac{B_l}{R^{l+2}}P_l(\cos\theta) - lA_l R^{l-1}P_l(\cos\theta).$$

Substitution of B_l yields

$$\left(\frac{\partial V_{\text{out}}}{\partial r} - \frac{\partial V_{\text{in}}}{\partial r}\right)\bigg|_{r=R} = \sum_{l=0}^{\infty} (2l+1)A_l R^{l-1}P_l(\cos\theta) = \frac{1}{\epsilon_0}\sigma_0(\theta).$$

Since Legendre polynomials are orthogonal, when $l \neq l'$ we have

$$\int_0^\pi P_l(\cos\theta)P_{l'}(\cos\theta)\sin\theta\, d\theta = \frac{2}{2l+1}.$$

It follows that

$$A_l = \frac{1}{2\epsilon_0 R^{l-1}}\int_0^\pi \sigma_0(\theta)P_l(\cos\theta)\sin\theta\, d\theta.$$

Note that $\sigma_0(\theta)$ can be rewritten as

$$\sigma_0(\theta) = \sin\theta\sin 3\theta = \sin\theta(\sin 2\theta\cos\theta + \cos 2\theta\sin\theta)$$

$$= \sin\theta\left[(2\sin\theta\cos\theta)\cos\theta + \sin\theta(2\cos^2\theta - 1)\right]$$

$$= 2\sin^2\theta\cos^2\theta + \sin^2\theta(2\cos^2\theta - 1)$$

$$= 2(1-\cos^2\theta)\cos^2\theta + (1-\cos^2\theta)(2\cos^2\theta - 1)$$

$$= 2\cos^2\theta - 2\cos^4\theta + 2\cos^2\theta - 2\cos^4\theta - 1 + \cos^2\theta$$

$$= -4\cos^4\theta + 5\cos^2\theta - 1.$$

We can find α, β, and γ such that
$$-4\cos^4\theta + 5\cos^2\theta - 1 = \alpha P_4(\cos\theta) + \beta P_2(\cos\theta) + \gamma P_0(\cos\theta).$$
So,
$$-4\cos^4\theta + 5\cos^2\theta - 1 = \alpha\frac{35\cos^4\theta - 30\cos^2\theta + 3}{8} + \beta\frac{3\cos^2\theta - 1}{2} + \gamma.$$

It follows that $\alpha = -\frac{32}{35}$, $\beta = \frac{22}{21}$, and $\gamma = -\frac{2}{15}$ and we can now solve for A_l.

$$A_l = \frac{1}{2\epsilon_0 R^{l-1}} \int_0^\pi \left[-\frac{32}{35} P_4(\cos\theta) P_l(\cos\theta)\sin\theta + \frac{22}{21} P_2(\cos\theta) P_l(\cos\theta)\sin\theta \right.$$
$$\left. -\frac{2}{15} P_0(\cos\theta) P_l(\cos\theta)\sin\theta \right] d\theta.$$

If $l = 4$, we have
$$A_4 = \frac{1}{2\epsilon_0 R^{4-1}} \int_0^\pi -\frac{32}{35}\left(\frac{35\cos^4\theta - 30\cos^2\theta + 3}{8}\right)^2 \sin\theta\, d\theta = -\frac{32}{315\epsilon_0 R^3}.$$

If $l = 2$, we have
$$A_2 = \frac{1}{2\epsilon_0 R^{2-1}} \int_0^\pi \frac{22}{21}\left(\frac{3\cos^2\theta - 1}{2}\right)^2 \sin\theta\, d\theta = \frac{22}{105\epsilon_0 R}.$$

If $l = 0$, we have
$$A_0 = \frac{1}{2\epsilon_0 R^{0-1}} \int_0^\pi -\frac{2}{15} \sin\theta\, d\theta = -\frac{2R}{15\epsilon_0}.$$

We can now find B_4, B_2, and B_0,
$$B_4 = A_4 R^{2(4)+1} = -\frac{32 R^6}{315\epsilon_0}$$

$$B_2 = A_2 R^{2(2)+1} = \frac{22 R^4}{105\epsilon_0}$$

$$B_0 = A_0 R^{2(0)+1} = -\frac{2R^2}{15\epsilon_0}.$$

Therefore, inside we have
$$V_{\text{in}}(r,\theta) = A_0 + A_2 r^2 P_2(\cos\theta) + A_4 r^4 P_4(\cos\theta).$$

So
$$V_{\text{in}}(r, \theta) = \frac{R}{\epsilon_0}\left[-\frac{2}{15} + \frac{22}{105\epsilon_0}\left(\frac{r}{R}\right)^2\left(\frac{3\cos^2\theta - 1}{2}\right)\right.$$
$$\left. - \frac{32}{315\epsilon_0}\left(\frac{r}{R}\right)^4\left(\frac{35\cos^4\theta - 30\cos^2\theta + 3}{8}\right)\right]$$

and outside we have
$$V_{\text{out}}(r, \theta) = \frac{B_0}{r^1} + \frac{B_2}{r^{2+1}}P_2(\cos\theta) + \frac{B_4}{r^{4+1}}P_4(\cos\theta).$$

So
$$V_{\text{out}}(r, \theta)$$
$$= \frac{R^2}{\epsilon_0 r}\left[-\frac{2}{15} + \frac{22}{105}\left(\frac{R}{r}\right)^2\left(\frac{3\cos^2\theta - 1}{2}\right) - \frac{32}{315}\left(\frac{R}{r}\right)^4\left(\frac{35\cos^4\theta - 30\cos^2\theta + 3}{8}\right)\right].$$

Problem 3.10. An infinitely long cylindrical shell of radius R is held at a potential $V_0(\phi) = \alpha \cos(4\phi)$. Find the potential inside and outside the shell.

Solution Our general solution in cylindrical coordinates is given by
$$V(s, \phi) = a_0 + b_0 \ln(s) + \sum_{k=1}^{\infty}\left\{s^k\left[a_k \cos(k\phi) + b_k \sin(k\phi)\right]\right.$$
$$\left. + s^{-k}\left[c_k \cos(k\phi) + d_k \sin(k\phi)\right]\right\}.$$

Inside:
Here, we must have $b_0 = c_k = d_k = 0$, otherwise the potential would blow up at the center. Our potential becomes
$$V(s, \phi) = a_0 + \sum_{k=1}^{\infty}s^k\left[a_k \cos(k\phi) + b_k \sin(k\phi)\right].$$

At the surface, we have
$$V(R, \phi) = a_0 + \sum_{k=1}^{\infty}R^k\left[a_k \cos(k\phi) + b_k \sin(k\phi)\right] = \alpha \cos(4\phi).$$

Note we have $a_0 = 0$, $b_k = 0$, and $a_k = 0$, except for a_4. So
$$R^4 a_4 \cos(4\phi) = \alpha \cos(4\phi) \rightarrow a_4 = \frac{\alpha}{R^4}.$$

Therefore, for $s \leq R$, we have
$$V(s, \phi) = \alpha\left(\frac{s}{R}\right)^4 \cos(4\phi).$$

Outside:
Here, we must have $b_0 = a_k = b_k = 0$, otherwise the potential would blow up as $s \to \infty$. Also, since we must have $V \to 0$ as $s \to \infty$, $a_0 = 0$. Our potential becomes

$$V(s, \phi) = \sum_{k=1}^{\infty} s^{-k}\Big[c_k \cos(k\phi) + d_k \sin(k\phi)\Big].$$

At the surface, we have

$$V(R, \phi) = \sum_{k=1}^{\infty} R^{-k}\Big[c_k \cos(k\phi) + d_k \sin(k\phi)\Big] = \alpha \cos(4\phi).$$

Note we have $d_k = 0$ and $c_k = 0$, except for c_4. So

$$R^{-4} c_4 \cos(4\phi) = \alpha \cos 4\phi \;\to\; c_4 = \alpha R^4.$$

Therefore, for $s \geqslant R$, we have

$$V(s, \phi) = \alpha \left(\frac{R}{s}\right)^4 \cos(4\phi).$$

Problem 3.11. Given an infinitely long cylindrical shell of radius R and surface charge $\sigma_0(\phi) = \alpha \cos(2\phi) + \beta \sin(3\phi)$, find the potential inside and outside the cylinder.

Solution Our general solution in cylindrical coordinates is given by

$$V(s, \phi) = a_0 + b_0 \ln(s) + \sum_{k=1}^{\infty} \Big\{s^k\Big[a_k \cos(k\phi) + b_k \sin(k\phi)\Big]$$

$$+ s^{-k}\Big[c_k \cos(k\phi) + d_k \sin(k\phi)\Big]\Big\}.$$

Inside, we must have $b_0 = c_k = d_k = 0$, otherwise the potential would blow up at the center. Our potential becomes

$$V_{\text{in}}(s, \phi) = a_0 + \sum_{k=1}^{\infty} s^k\Big[a_k \cos(k\phi) + b_k \sin(k\phi)\Big].$$

Outside we must have $b_0 = a_k = b_k = 0$, otherwise the potential would blow up as $s \to \infty$. Also, since we must have $V \to 0$ as $s \to \infty$, $a_0 = 0$. Our potential becomes

$$V_{\text{out}}(s, \phi) = \sum_{k=1}^{\infty} s^{-k}\Big[c_k \cos(k\phi) + d_k \sin(k\phi)\Big].$$

At the surface, we have

$$\left(\frac{\partial V_{\text{out}}}{\partial s} - \frac{\partial V_{\text{in}}}{\partial s}\right)\bigg|_{s=R} = -\frac{1}{\epsilon_0}\sigma_0(\phi),$$

3-23

where

$$\frac{\partial V_{\text{out}}}{\partial s} = \sum_{k=1}^{\infty} -ks^{-k-1}\left[c_k \cos(k\phi) + d_k \sin(k\phi)\right]$$

and

$$\frac{\partial V_{\text{in}}}{\partial s} = \sum_{k=1}^{\infty} ks^{k-1}\left[a_k \cos(k\phi) + b_k \sin(k\phi)\right].$$

Thus,

$$\left(\frac{\partial V_{\text{out}}}{\partial s} - \frac{\partial V_{\text{in}}}{\partial s}\right)\bigg|_{s=R} = \sum_{k=1}^{\infty} -kR^{-k-1}\left[c_k \cos(k\phi) + d_k \sin(k\phi)\right]$$

$$- kR^{k-1}\left[a_k \cos(k\phi) + b_k \sin(k\phi)\right]$$

$$= -\frac{\alpha \cos(2\phi) + \beta \sin(3\phi)}{\epsilon_0}.$$

From this, we can see that $c_k = a_k = 0$, except when $k = 2$, and $d_k = b_k = 0$, except when $k = 3$. This means

$$2\cos(2\phi)\left(R^{-3}c_2 + Ra_2\right) + 3\sin(3\phi)\left(R^{-4}d_3 + R^2 b_3\right) = \frac{\alpha \cos(2\phi)}{\epsilon_0} + \frac{\beta \sin(3\phi)}{\epsilon_0}.$$

Separating out the sine and cosine term, we have

$$2\left(R^{-3}c_2 + Ra_2\right) = \frac{\alpha}{\epsilon_0} \rightarrow c_2 = R^3\left(\frac{\alpha}{2\epsilon_0} - Ra_2\right)$$

and

$$3(R^{-4}d_3 + R^2 b_3) = \frac{\beta}{\epsilon_0} \rightarrow d_3 = R^4\left(\frac{\beta}{3\epsilon_0} - R^2 b_3\right).$$

Since V is continuous, we have

$$V_{\text{out}}(R, \phi) = V_{\text{in}}(R, \phi)$$

$$R^{-2}c_2 \cos(2\phi) + R^{-3}d_3 \sin(3\phi) = a_0 + R^2 a_2 \cos(2\phi) + R^3 b_3 \sin(3\phi).$$

We can see that $a_0 = 0$. Also, considering the sine and cosine terms separately, we have

$$R^{-2}c_2 = R^2 a_2.$$

Substitution of c_2 yields

$$R^{-2}R^3\left(\frac{\alpha}{2\epsilon_0} - Ra_2\right) = R^2 a_2$$

so

$$a_2 = \frac{\alpha}{4R\epsilon_0}.$$

Also

$$R^{-3}d_3 = R^3 b_3.$$

Substitution of d_3 yields

$$R^{-3}R^4\left(\frac{\beta}{3\epsilon_0} - R^2 b_3\right) = R^3 b_3$$

So

$$b_3 = \frac{\beta}{6R^2\epsilon_0}.$$

Therefore,

$$c_2 = R^3\left(\frac{\alpha}{2\epsilon_0} - R\frac{\alpha}{4R\epsilon_0}\right) = \frac{\alpha R^3}{4\epsilon_0}$$

and

$$d_3 = R^4\left(\frac{\beta}{3\epsilon_0} - R^2\frac{\beta}{6R^2\epsilon_0}\right) = \frac{\beta R^4}{6\epsilon_0}.$$

Combining everything, the potential inside is

$$V_{\text{in}}(s,\phi) = s^2\frac{\alpha}{4R\epsilon_0}\cos(2\phi) + s^3\frac{\beta}{6R^2\epsilon_0}\sin(3\phi)$$

$$= \frac{R}{\epsilon_0}\left[\frac{\alpha}{4}\left(\frac{s}{R}\right)^2\cos(2\phi) + \frac{\beta}{6}\left(\frac{s}{R}\right)^3\sin(3\phi)\right]$$

and outside is

$$V_{\text{out}}(s,\phi) = s^{-2}\frac{\alpha R^3}{4\epsilon_0}\cos(2\phi) + s^{-3}\frac{\beta R^4}{6\epsilon_0}\sin(3\phi)$$

$$= \frac{R}{\epsilon_0}\left[\frac{\alpha}{4}\left(\frac{R}{s}\right)^2\cos(2\phi) + \frac{\beta}{6}\left(\frac{R}{s}\right)^3\sin(3\phi)\right].$$

Problem 3.12. The electric potential varies as $\frac{1}{r}$ for a monopole, as $\frac{1}{r^2}$ for a dipole, as $\frac{1}{r^3}$ for a quadrupole, and as $\frac{1}{r^4}$ for an octopole. How will the electric potential depend on r for a mutipole with n charges (n being a k power of 2, $n = 2^k$)?

Solution

	Number of charges	Potential
Monopole	$n = 2^0 = 1; k = 0$	$V \sim \frac{1}{r^{k+1}} = \frac{1}{r}$
Dipole	$n = 2^1 = 2; k = 1$	$V \sim \frac{1}{r^{k+1}} = \frac{1}{r^2}$
Quadrupole	$n = 2^2 = 4; k = 2$	$V \sim \frac{1}{r^{k+1}} = \frac{1}{r^3}$
Octopole	$n = 2^3 = 8; k = 3$	$V \sim \frac{1}{r^{k+1}} = \frac{1}{r^4}$
Multipole	$n = 2^k; k$	$V \sim \frac{1}{r^{k+1}}$

Problem 3.13. Let us consider an electric dipole with charges q and $-q$ situated at distance d from each other, shown below. Calculate the electric potential at a point P in the far approximation $r \gg d$.

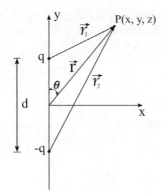

Solution The total electric potential is obtained by superposition

$$V(\vec{r}) = V_1 + V_2 = \frac{1}{4\pi\varepsilon_0}\left(\frac{q}{r_1} + \frac{-q}{r_2}\right).$$

From the law of cosines

$$r_1^2 = \left(\frac{d}{2}\right)^2 + r^2 - 2\frac{d}{2}r\cos\theta$$

and

$$r_2^2 = \left(\frac{d}{2}\right)^2 + r^2 - 2\frac{d}{2}r\cos(\pi - \theta).$$

Note that $\cos(\pi - \theta) = -\cos\theta$. We can rewrite our potential as

$$V = \frac{q}{4\pi\varepsilon_0}\left(\frac{1}{\sqrt{\frac{d^2}{4} + r^2 - rd\cos\theta}} - \frac{1}{\sqrt{\frac{d^2}{4} + r^2 + rd\cos\theta}}\right)$$

or

$$V = \frac{q}{4\pi\varepsilon_0}\left(\frac{1}{r\sqrt{1 + \frac{d^2}{4r^2} - \frac{d}{r}\cos\theta}} - \frac{1}{r\sqrt{1 + \frac{d^2}{4r^2} + \frac{d}{r}\cos\theta}}\right).$$

When $r \gg d$, $\frac{d^2}{4r^2}$ is very small and can be ignored. If we consider $x = \frac{d}{r}\cos\theta \ll 1$, we can use the binomial theorem and obtain

$$(1 + x)^{-\frac{1}{2}} \cong 1 - \frac{x}{2}$$

and

$$(1 - x)^{-\frac{1}{2}} \cong 1 + \frac{x}{2}.$$

From this, we have

$$\frac{1}{r\sqrt{1 - \frac{d}{r}\cos\theta}} = \frac{1}{r}\left(1 + \frac{d}{2r}\cos\theta\right)$$

and

$$\frac{1}{r\sqrt{1 + \frac{d}{r}\cos\theta}} = \frac{1}{r}\left(1 - \frac{d}{2r}\cos\theta\right).$$

Therefore,

$$V = \frac{q}{4\pi\varepsilon_0}\left[\frac{1}{r}\left(1 + \frac{d}{2r}\cos\theta\right) - \frac{1}{r}\left(1 - \frac{d}{2r}\cos\theta\right)\right]$$

$$= \frac{q}{4\pi\varepsilon_0 r}\left[1 + \frac{d}{2r}\cos\theta - 1 + \frac{d}{2r}\cos\theta\right]$$

$$V = \frac{qd}{4\pi\varepsilon_0 r^2}\cos\theta.$$

Taking the dipole moment $\vec{p} = q\vec{d}$, we have

$$V = \frac{\vec{p}\cdot\hat{r}}{4\pi\varepsilon_0 r^2}.$$

Problem 3.14. Find the electric field of the dipole in problem 3.13, centered at the origin with the dipole moment $\vec{p}$ in the z-direction.

Solution From problem 3.13, the electric potential is given by

$$V(\vec{r}) = \frac{qd}{4\pi\varepsilon_0 r^2} \cos\theta = \frac{\vec{p}\cdot\hat{r}}{4\pi\varepsilon_0 r^2} = \frac{p\cos\theta}{4\pi\varepsilon_0 r^2}.$$

We can find the field from the potential using

$$\vec{E} = -\nabla V.$$

We need to use the gradient in spherical coordinates

$$E_r = -\frac{\partial V}{\partial r} = \frac{2p\cos\theta}{4\pi\varepsilon_0 r^3}$$

$$E_\theta = -\frac{1}{r}\frac{\partial V}{\partial \theta} = \frac{p\sin\theta}{4\pi\varepsilon_0 r^3}$$

$$E_\phi = -\frac{1}{r\sin\theta}\frac{\partial V}{\partial \phi} = 0.$$

Therefore, the electric field due to the dipole is

$$\vec{E}_{\text{dipole}}(r, \theta) = \frac{p}{4\pi\varepsilon_0 r^3}\left(2\cos\theta\,\hat{r} + \sin\theta\,\hat{\theta}\right).$$

Problem 3.15. Two point charges $+4q$ and $-q$ are separated by a distance d. The first charge is placed at $(0, 0, d)$ and the second one at the origin. Find: (a) the monopole moment; (b) the dipole moment; (c) the electric potential in spherical coordinates for $r \gg d$. Include only the monopole and dipole contributions.

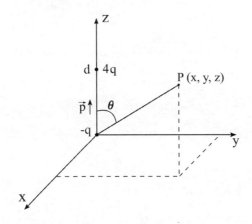

Solution
(a) Monopole moment:
$$Q = 4q - q = 3q.$$

(b) Dipole moment:
$$\vec{p} = \sum_{i=1}^{2} q_i \vec{r}_i = -q(0, 0, 0) + 4q(0, 0, d) = 4qd\hat{z}.$$

(c) The electric potential:
$$V(\vec{r}) = \frac{1}{4\pi\varepsilon_0} \left(\frac{1}{r} \sum_i q_i + \frac{1}{r^2} \sum_i q_i r_i' \cos\theta_i' + \cdots \right) = \frac{1}{4\pi\varepsilon_0} \left(\frac{3q}{r} + \frac{1}{r^2} \vec{p} \cdot \hat{r} \right)$$

$$= \frac{1}{4\pi\varepsilon_0} \left(\frac{3q}{r} + \frac{p \cos\theta}{r^2} \right) = \frac{1}{4\pi\varepsilon_0} \left(\frac{3q}{r} + \frac{4qd \cos\theta}{r^2} \right).$$

Bibliography

Byron F W and Fuller R W 1992 *Mathematics of Classical and Quantum Physics* (New York: Dover)
Griffiths D J 1999 *Introduction to Electrodynamics* 3rd edn (Englewood Cliffs, NJ: Prentice Hall)
Griffiths D J 2013 *Introduction to Electrodynamics* 4th edn (New York: Pearson)
Halliday D, Resnick R and Walker J 2010 *Fundamentals of Physics* 9th edn (New York: Wiley)
Halliday D, Resnick R and Walker J 2013 *Fundamentals of Physics* 10th edn (New York: Wiley)
Jackson J D 1998 *Classical Electrodynamics* 3rd edn (New York: Wiley)
Rogawski J 2011 *Calculus: Early Transcendentals* 2nd edn (San Francisco, CA: Freeman)

IOP Concise Physics

Electromagnetism
Problems and solutions

Carolina C Ilie and Zachariah S Schrecengost

Chapter 4

Magnetostatics

This chapter introduces magnetic fields in a vacuum and the methods for calculating the magnetic field. Magnetic fields are intrinsically determined by electric charges in motion. We imagine these small currents as magnetic dipoles. From the general Biot–Savart law, to the more straightforward Ampère's law applicable to configurations with higher degree of symmetry, the suggested problems constitute good practice in magnetostatics.

4.1 Theory

4.1.1 Magnetic force

A charge q moving with velocity $\vec{v}$ in a magnetic field $\vec{B}$ experiences a force given by

$$\vec{F}_m = q\vec{v} \times \vec{B}.$$

4.1.2 Force on a current carrying wire

The force on a current carrying wire in a magnetic field $\vec{B}$ is

$$\vec{F}_m = \int I(d\vec{\ell} \times \vec{B}).$$

4.1.3 Volume current density

The current density of a current $\vec{I}$ is

$$\vec{J} = \frac{d\vec{I}}{da_\perp}$$

and the current density of a charge density ρ moving at velocity $\vec{v}$ is

$$\vec{J} = \rho\vec{v}.$$

4.1.4 Continuity equation

The divergence of the charge density $\vec{J}$ is related to the charge density ρ by

$$\nabla \cdot \vec{J} = -\frac{\partial \rho}{\partial t}.$$

4.1.5 Biot–Savart law

The magnetic field due to current distributions is given by

$$\vec{B}(\vec{r}) = \frac{\mu_0 I}{4\pi} \int \frac{d\vec{\ell}' \times \hat{r}}{r^2}$$

$$\vec{B}(\vec{r}) = \frac{\mu_0}{4\pi} \int \frac{\vec{K}(\vec{r}') \times \hat{r}}{r^2} da'$$

$$\vec{B}(\vec{r}) = \frac{\mu_0}{4\pi} \int \frac{\vec{J}(\vec{r}') \times \hat{r}}{r^2} d\tau'.$$

4.1.6 Divergence of $\vec{B}$

Given magnetic field $\vec{B}$, we have

$$\nabla \cdot \vec{B} = 0.$$

4.1.7 Ampère's law

Given magnetic field $\vec{B}$, we have

$$\nabla \times \vec{B} = \mu_0 \vec{J}.$$

By applying Stoke's law, we also have

$$\oint_S \vec{B} \cdot d\vec{\ell} = \mu_0 I_{\text{enc}},$$

where

$$I_{\text{enc}} = \int \vec{J} \cdot d\vec{a}.$$

4.1.8 Vector potential

The vector potential due to current distributions is given by

$$\vec{A}(\vec{r}) = \frac{\mu_0 I}{4\pi} \int \frac{1}{r} d\vec{\ell}'$$

$$\vec{A}(\vec{r}) = \frac{\mu_0}{4\pi} \int \frac{\vec{K}(\vec{r}')}{r} da'$$

$$\vec{A}(\vec{r}) = \frac{\mu_0}{4\pi} \int \frac{\vec{J}(\vec{r}')}{r} d\tau'.$$

Also,
$$\vec{B} = \nabla \times \vec{A}$$
and
$$\nabla^2 \vec{A} = \mu_0 \vec{J}.$$

4.1.9 Magnetic dipole moment
The magnetic dipole moment due to a current I is
$$\vec{m} = \int I \, d\vec{a}.$$

4.1.10 Magnetic field due to dipole moment
Given magnetic dipole moment $\vec{m}$, the magnetic field is
$$\vec{B}_{\text{dip}} = \frac{\mu_0 m}{4\pi r^3} \left(2\cos\theta \, \hat{r} + \sin\theta \, \hat{\theta} \right).$$

4.2 Problems and solutions

Problem 4.1. A proton travels through a uniform magnetic and electric field. The magnetic field is $\vec{B} = a\hat{y}$, where a is a positive constant. If at one moment the velocity of the proton is $\vec{v} = b\hat{z}$, where b is a positive constant, what is the force acting on the proton if the electric field is $\vec{E} = -c\hat{x}$?

Solution

$$\vec{F} = q(\vec{E} + \vec{v} \times \vec{B}) = q\left[-c\hat{x} + (b\hat{z}) \times (a\hat{y})\right] = q\left[-c\hat{x} + ab(-\hat{x})\right] = (c\hat{x} + ab\hat{x})$$

$$\vec{F} = -q(c + ab)\hat{x}.$$

Problem 4.2. A particle of charge q enters a region of uniform magnetic field $\vec{B}$ (out of the page, in the z-direction) with an initial velocity $\vec{v}$ (in the x-direction). The particle is deflected a distance y above the initial direction. If the region has a width of x, find the sign of the charge and the deflected distance y as a function of q, v, B, and x.

Solution

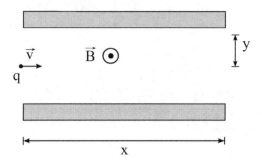

$$\vec{F} = q\vec{v} \times \vec{B}$$

Since the charge is deflected as shown, the charge is negative (determined from the right-hand rule). In the x-direction we have no force, and therefore no acceleration

$$x = vt.$$

In the y-direction

$$ma = |q|vB \sin 90.$$

So

$$a = \frac{|q|vB}{m}$$

and

$$y = y_o + v_{oy}t + \frac{a_y t^2}{2} = \frac{a_y t^2}{2} = \frac{|q|vBt^2}{2m}.$$

By substituting the time

$$t = \frac{x}{v}$$

we obtain

$$y = \frac{|q|vB\frac{x^2}{v^2}}{2m} = \frac{|q|Bx^2}{2mv} = \frac{|q|Bx^2}{2p},$$

where p is the momentum of the particle.

Problem 4.3. The current density in a wire of circular cross section of radius R is dependent on the distance from the axis, given by $\vec{J} = ks^2 \hat{z}$, where k is a constant. Find a) the total current in the wire and b) the current density if the current in a) is uniformly distributed.

Solution

a) Given current density $\vec{J} = ks^2\hat{z}$, the current is

$$I = \int \vec{J} \cdot d\vec{a} = \int \left(ks^2\hat{z}\right) \cdot \left(s\, d\phi\, ds\, \hat{z}\right) = \int_0^{2\pi} d\phi \int_0^R ks^3 ds = 2\pi k \left.\frac{s^4}{4}\right|_0^R = \frac{\pi k R^4}{2}.$$

b) If this current was uniformly distributed, the current density is simply

$$J = \frac{I}{\text{area}} = \frac{1}{\pi R^2} \frac{\pi k R^4}{2} = \frac{k R^2}{2}.$$

Problem 4.4.

a) In the famous experiment of J J Thompson, he measured the charge to mass $\frac{q}{m}$ ratio of the catode rays. Find $\frac{q}{m}$ when you know B, R, and v, and that $\vec{B}$ is perpendicular to $\vec{v}$.

b) He also had the beams going in a region with perpendicular electric field and magnetic field and 'tuned' them such that the electrons left the region with unchanged direction. If the speed of the electrons is v and the magnetic field is $\vec{B}$, what should be the value of the electric field?

Solution

a) The magnitude of the magnetic force is given by

$$\left|\vec{F}_m\right| = |q|\vec{v} \times \vec{B},$$

where we have $\vec{v} \perp \vec{B}$. Also

$$\vec{F}_m = \vec{F}_{\text{centripetal}}.$$

So,

$$F_m = |q|vB$$

and

$$F_{\text{cent}} = \frac{mv^2}{R}.$$

Therefore from

$$|q|vB = \frac{mv^2}{R}$$

we have

$$\frac{|q|}{m} = \frac{v}{BR}.$$

b) Setting the magnetic and electric forces as equal, we have

$$\vec{F}_m = \vec{F}_e \rightarrow q\vec{v} \times \vec{B} = q\vec{E}.$$

Dividing by q and expressing this in terms of magnitudes, we have

$$vB \sin 90 = E$$

so

$$E = vB.$$

Problem 4.5. Find the magnetic field at:
a) The center of a circular wire loop of radius R carrying current I.

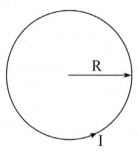

b) The center of a wire loop that consists of half a loop of radius R and half a square loop of side $2R$, carrying current I.

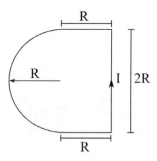

Solution
a) The Biot–Savart law states

$$\vec{B} = \frac{\mu_0 I}{4\pi} \int \frac{d\vec{\ell} \times \hat{r}}{r^2},$$

where

$$d\vec{\ell} = R\,d\vec{\phi} = R\,d\phi\,\hat{\phi}$$

and

$$\hat{r} = -\hat{r}.$$

Since $r = R$, the Biot–Savart law becomes

$$\vec{B} = \frac{\mu_0 IR}{4\pi R^2} \int_0^{2\pi} \hat{\phi} \times \hat{r}\,d\phi = \frac{\mu_0 I}{2R}\hat{z} \quad \text{(out of page)}.$$

b) From part a), we can determine the field contribution due to the circular part is

$$\vec{B}_c = \frac{\mu_0 I}{4R}\hat{z},$$

which is half that of the full loop. As for the square, we consider the field R above the wire. We have

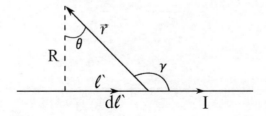

So

$$\vec{B} = \frac{\mu_0 I}{4\pi} \int \frac{d\vec{\ell}' \times \hat{r}}{r^2},$$

where $d\vec{\ell}' \times \hat{r}$ points in the $\hat{z}$-direction (out of the page). Also,

$$d\ell' \sin\gamma = d\ell' \cos\theta$$

and

$$\ell' = R\tan\theta \to d\ell' = \frac{R}{\cos^2\theta}d\theta.$$

and
$$r^2 = \ell'^2 + R^2 \rightarrow \frac{1}{r^2} = \frac{\cos^2\theta}{R^2}.$$

Therefore,
$$\vec{B} = \frac{\mu_0 I}{4\pi} \int_{\theta_1}^{\theta_2} \left(\frac{\cos^2\theta}{R^2}\right)\left(\frac{R}{\cos^2\theta}\right)\cos\theta\, d\theta = \frac{\mu_0 I}{4\pi R}(\sin\theta_2 - \sin\theta_1).$$

So for each R-lengthed segment (i) and (ii), we have $\theta_1 = 0$ and $\theta_2 = \frac{\pi}{4}$, and for the $2R$-lengthed segment, we have $\theta_1 = -\frac{\pi}{4}$ and $\theta_2 = \frac{\pi}{4}$. So

$$\vec{B} = \vec{B}_c + 2\left(\frac{\mu_0 I}{4\pi R}\sin\frac{\pi}{4}\hat{z}\right) + \frac{\mu_0 I}{4\pi R}\left[\sin\left(\frac{\pi}{4}\right) - \sin\left(-\frac{\pi}{4}\right)\right]\hat{z}$$
$$= \frac{\mu_0 I}{4\pi R}\left(\pi + 2\sqrt{2}\right)\hat{z} \quad \text{(out of page)}.$$

Problem 4.6. Consider a cylindrical shell of radius R and length L, carrying σ and rotating at ω. Find the magnetic field d from the end of the shell (on the axis).

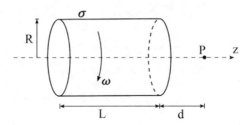

Solution Here we have
$$\vec{B} = \frac{\mu_0}{4\pi}\int\frac{\vec{K}\times\hat{r}}{r^2}da,$$
where
$$da = R\,dz\,d\phi$$
and $0 \leqslant z \leqslant L$. The surface charge is given by
$$\vec{K} = \sigma\vec{v} = \sigma\omega R\hat{\phi}.$$

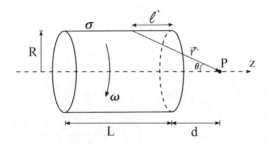

From the figure above, we have $\ell' = L - z$ and

$$r = \sqrt{(\ell' + d)^2 + R^2} = \sqrt{(L - z + d)^2 + R^2}.$$

Note that the field cancels such that the $\hat{z}$-component is the only component that survives. So

$$\left[\vec{K} \times \hat{r}\right]_z = \sigma \omega R \sin\theta \, \hat{z} = \sigma \omega R \frac{R}{r} \hat{z} = \frac{\sigma \omega R^2}{r} \hat{z}.$$

Putting everything together, we have

$$\vec{B} = \frac{\mu_0}{4\pi} \int_0^{2\pi} \int_0^L \frac{\sigma \omega R^2 R}{\left[(L - z + d)^2 + R^2\right]^{3/2}} \hat{z} \, dz \, d\phi$$

$$= \frac{\mu_0 \sigma \omega R^3 \hat{z}}{2} \int_0^L \frac{dz}{\left[(L - z + d)^2 + R^2\right]^{3/2}}.$$

Therefore,

$$\vec{B} = \frac{\mu_0 \sigma \omega R}{2} \left[\frac{d + L}{\sqrt{R^2 + (d + L)^2}} - \frac{d}{\sqrt{R^2 + d^2}}\right] \hat{z}.$$

Problem 4.7. A hemisphere of radius R and charge density ρ is rotating at ω. Find the magnetic field d above the center.

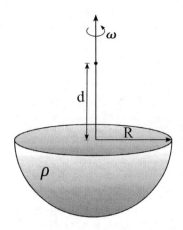

Solution Here we have

$$\vec{B} = \frac{\mu_0}{4\pi} \int \frac{\vec{J} \times \hat{r}}{r^2} d\tau,$$

where

$$d\tau = r^2 \sin\theta \, dr \, d\phi \, d\theta.$$

From the figure below, we can see that

$$\tilde{r}^2 = d^2 + r^2 - 2dr \cos\theta.$$

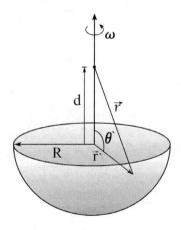

Also,

$$\vec{J} = \rho\vec{v} = \rho\omega r \sin\theta \, \hat{\phi}.$$

Note that the field cancels such that the $\hat{z}$-component is the only component that survives. So
$$\left[\vec{J} \times \hat{r}\right]_z = \rho\omega r \sin\theta \left(\frac{r\sin\theta}{r}\right)\hat{z}.$$

Therefore
$$\vec{B} = \frac{\mu_0}{4\pi} \int_0^{2\pi} \int_{\frac{\pi}{2}}^{\pi} \int_0^R \frac{\rho\omega r^2 \sin^2\theta \, r^2 \sin\theta \, \hat{z}}{\left(d^2 + r^2 - 2dr\cos\theta\right)^{\frac{3}{2}}} dr \, d\theta \, d\phi$$

$$= \frac{\mu_0 \rho \omega}{2} \int_{\frac{\pi}{2}}^{\pi} \int_0^R \frac{r^4 \sin^3\theta \, \hat{z}}{\left(d^2 + r^2 - 2dr\cos\theta\right)^{\frac{3}{2}}} dr \, d\theta$$

$$\vec{B} = \frac{\mu_0 \rho \omega}{30 d^3}\left[\sqrt{R^2 + d^2}\left(-2R^4 + d^2 R^2 - 12 d^4\right) + 2R^5 + 5d^3 R^2 + 12 d^5\right]\hat{z}.$$

Problem 4.8. A spherical shell of radius R, carrying σ and rotating at ω, is centered at the origin. Find the velocity a loop of wire, carrying λ with radius a centered at the origin, required to cancel the magnetic field at the center.

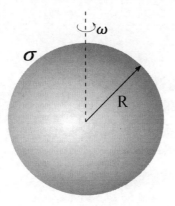

Solution First, we have
$$\vec{B} = \frac{\mu_0}{4\pi} \int \frac{\vec{K} \times \hat{r}}{r^2} da,$$

where
$$r = R$$

4-11

$$\hat{r}' = -\hat{r}$$

$$\vec{K} = \sigma \vec{v} = \sigma \omega R \sin\theta \, \hat{\phi}$$

$$da = R^2 \sin\theta \, d\theta \, d\phi$$

and

$$\vec{K} \times \hat{r}' = \sigma \omega R \sin^2\theta \, \hat{z}.$$

Putting this together, we have

$$\vec{B}_s = \frac{\mu_0}{4\pi} \int_0^\pi \int_0^{2\pi} \frac{\sigma \omega R \sin^2\theta \, R^2 \sin\theta \, \hat{z}}{R^2} d\phi \, d\theta = \frac{\mu_0 \sigma \omega R \, \hat{z}}{2} \int_0^\pi \sin^3\theta \, d\theta$$

$$\vec{B}_s = \frac{2\mu_0 \sigma \omega R}{3} \hat{z}.$$

Now a line of charge λ rotating at $\vec{v}$ 'looks' like a wire carrying current $\vec{I} = \lambda\vec{v} = \lambda\omega_l a\hat{\phi}$. From problem 4.5(a), we know this produces magnetic field

$$\vec{B}_l = \frac{\mu_0 \lambda \omega_l a}{2a} \hat{z} = \frac{\mu_0 \lambda \omega_l}{2} \hat{z}.$$

We want $\vec{B}_l + \vec{B}_s = 0$, so

$$\frac{\mu_0 \lambda \omega_l}{2} + \frac{2\mu_0 \sigma \omega R}{3} = 0 \rightarrow \frac{\lambda \omega_l}{2} = -\frac{2\sigma \omega R}{3}.$$

Therefore,

$$\omega_l = -\frac{4\omega \sigma R}{3\lambda}.$$

Problem 4.9. A long straight wire carries a steady current I. Obtain the magnetic field at a distance s from the wire.

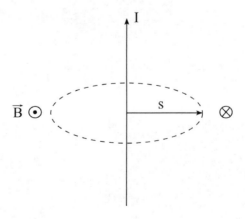

Solution Here, we apply Ampère's law, with an Amperian loop that is a circle centered on the wire and in a plane perpendicular to the wire. Ampère's law is

$$\oint_S \vec{B} \cdot d\vec{\ell} = \mu_0 I_{\text{enc}}.$$

Noticing $\vec{B} \parallel d\vec{\ell}$, it follows that B is a constant at a certain distance s from the wire. So the left-hand side is given by

$$\oint_S \vec{B} \cdot d\vec{\ell} = \oint_S B \, d\ell = B \oint_S d\ell = B 2\pi s.$$

The enclosed current is just simply given by

$$I_{\text{enc}} = I$$

So

$$B = \frac{\mu_0 I}{2\pi s}.$$

Since the magnetic field is tangent on the circle at every point

$$\vec{B} = \frac{\mu_0 I}{2\pi s} \hat{\phi}.$$

Problem 4.10. An electric current flows through a long cylinder wire of radius a. Find the magnetic field inside and outside the wire, and plot it, in the following cases, where k is a constant with the appropriate units:
a) $I = $ constant (steady current).
b) Current density J is proportional to the distance from the axis: $J = ks$.
c) $J = ks^2$.

Solution
a) Here we have constant I. For $s > a$, our Amperian loop is given by

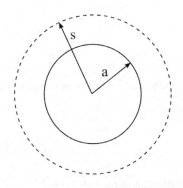

Ampère's law states
$$\oint_S \vec{B} \cdot d\vec{\ell} = \mu_0 I_{\text{enc}}$$

In all cases, the left-hand side yields
$$\oint_S \vec{B} \cdot d\vec{\ell} = \oint_S B \, d\ell = B \oint_S d\ell = B 2\pi s.$$

Here we simply have
$$I_{\text{enc}} = I$$

So
$$B 2\pi s = \mu_0 I \rightarrow B = \frac{\mu_0 I}{2\pi s}.$$

Applying the right-hand rule to a current coming out of the page, we have
$$\vec{B} = \frac{\mu_0 I}{2\pi s} \hat{\phi}.$$

For $s < a$, our Amperian loop is inside the wire, at radius s.

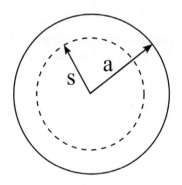

Again we have
$$\oint_S \vec{B} \cdot d\vec{\ell} = \mu_0 I_{\text{enc}}.$$

Since I is uniform, the current density is constant,
$$J = \frac{I}{\pi a^2} = \frac{I_{\text{enc}}}{\pi s^2}$$

so
$$I_{\text{enc}} = I \frac{s^2}{a^2}.$$

Therefore

$$B = \frac{\mu_0 s I}{2\pi a^2} \rightarrow \vec{B} = \frac{\mu_0 s I}{2\pi a^2}\hat{\phi}.$$

The plot of the magnetic field is given below.

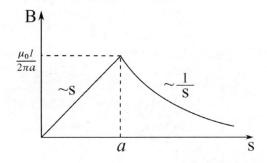

b) Here we have $J = ks$. For $s > a$, the Amperian loop is the same as part a) for $s > a$. We still have the left hand side of Ampère's law given by

$$\oint_S \vec{B} \cdot d\vec{\ell} = B 2\pi s.$$

Now, since J is not constant, we need to integrate in order to find the enclosed current, so

$$I_{\text{enc}} = \int J\, da = \int_0^a Js\, ds \int_0^{2\pi} d\phi = \int_0^a ks^2\, ds \int_0^{2\pi} d\phi = 2\pi \frac{ks^3}{3}\bigg|_0^a = \frac{2\pi k a^3}{3}.$$

Therefore,

$$B 2\pi s = \frac{\mu_0 2\pi k a^3}{3} \rightarrow B = \frac{\mu_0 k a^3}{3s}$$

with

$$\vec{B} = \frac{\mu_0 k a^3}{3s}\hat{\phi}.$$

For $s < a$, we again have the same Amperian loop as part a). Here, our enclosed current is given by

$$I_{\text{enc}} = \int J\, da = \int_0^s ks'\, s'\, ds' \int_0^{2\pi} d\phi = 2\pi \frac{ks'^3}{3}\bigg|_0^s = \frac{2\pi k s^3}{3}.$$

Therefore,

$$B2\pi s = \frac{\mu_0 2\pi k s^3}{3} \to B = \frac{\mu_0 k s^2}{3}$$

with

$$\vec{B} = \frac{\mu_0 k s^2}{3}\hat{\phi}.$$

The plot of the magnetic field is given below.

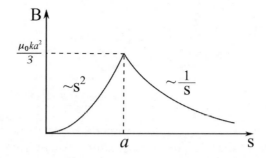

c) Now we have $J = ks^2$. For $s > a$, the enclosed current is given by

$$I_{\text{enc}} = \int J\, da = \int_0^a Js\, ds \int_0^{2\pi} d\phi = 2\pi \int_0^a ks^2\, s\, ds = 2\pi \frac{ks^4}{4}\bigg|_0^a = 2\pi \frac{ka^4}{4}.$$

Therefore,

$$B2\pi s = \mu_0 2\pi \frac{ka^4}{4} \to B = \frac{\mu_0 ka^4}{4s}$$

with

$$\vec{B} = \frac{\mu_0 ka^4}{4s}\hat{\phi}.$$

For $s < a$, we have

$$I_{\text{enc}} = \int J\, da = \int_0^s Js'\, ds' \int_0^{2\pi} d\phi = 2\pi \int_0^s ks'^2\, s'\, ds'' = \frac{2\pi ks'^4}{4}\bigg|_0^s = \frac{2\pi ks^4}{4}.$$

Therefore,

$$B = \frac{\mu_0 ks^3}{4}$$

4-16

with

$$\vec{B} = \frac{\mu_0 k s^3}{4} \hat{\phi}.$$

The plot of magnetic field is given below.

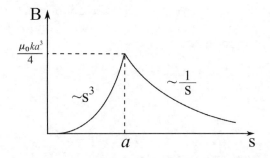

Problem 4.11. Use Ampère's law to obtain the magnetic field inside and outside a solenoid of $n = \frac{N}{L}$, where N is the number of turns, and L is the length of the solenoid. The solenoid is carrying the current I.

Solution Let us choose two Amperian loops given by

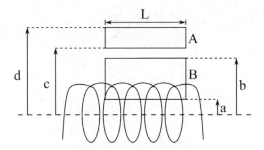

Starting with the outside loop (loop A), the magnetic field does not have any radial B_r component or B_ϕ. Ampère's law is given by

$$\oint_S \vec{B} \cdot d\vec{\ell} = \mu_0 I_{\text{enc}}.$$

Since we have $I_{enc} = 0$, $B(c) = B(d)$, but since $B \to 0$ for large distances, $B = 0$ outside the solenoid. For loop B, the left hand side of Ampère's law is given by

$$\oint_S \vec{B} \cdot d\vec{\ell} = \oint_S B \, d\ell = BL.$$

The sides perpendicular on the solenoid yield zero dot product, as the magnetic field is oriented parallel to the solenoid's axis in the z-direction (by the right-hand rule). The enclosed current is given by

$$I_{enc} = InL = IN.$$

Substituting these quantities into Ampère's law yields

$$BL = \mu_0 InL \to B = \mu_0 In = \frac{\mu_0 IN}{L}$$

with

$$\vec{B} = \mu_0 In\hat{z} = \frac{\mu_0 IN}{L}\hat{z}.$$

Problem 4.12. A current carrying empty cylinder of inner radius a and outer radius b has a current density J, which is proportional to the distance from the axis; $J = ks$, k constant. Find the magnetic field in all regions.

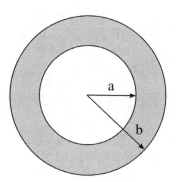

Solution There are three significant regions: $s < a$, $a < s < b$, and $s > b$. The easiest to find is the field for $s < a$,

4-18

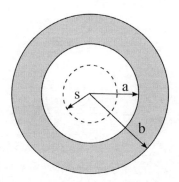

where the enclosed current is zero. Therefore,

$$\oint_S \vec{B} \cdot d\vec{\ell} = 0,$$

so $\vec{B} = 0$. For $a < s < b$, we have

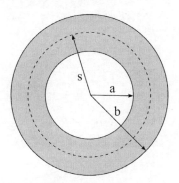

Using Ampère's law,

$$\oint_S \vec{B} \cdot d\vec{\ell} = \mu_0 I_{\text{enc}}.$$

We have the left-hand side is given by

$$\oint_S \vec{B} \cdot d\vec{\ell} = B 2\pi s,$$

with enclosed current given by

$$I_{\text{enc}} = \int \vec{J} \cdot d\vec{a} = \int_a^s J s' \, ds' \int_0^{2\pi} d\phi = 2\pi \int_a^s k s' \, s' \, ds' = \frac{2\pi k s'^3}{3} \bigg|_a^s = \frac{2\pi k (s^3 - a^3)}{3}$$

$$B = \frac{\mu_o k(s^3 - a^3)}{3s}$$

with

$$\vec{B} = \frac{\mu_o k(s^3 - a^3)}{3s}\hat{\phi}.$$

For $s > b$, we have

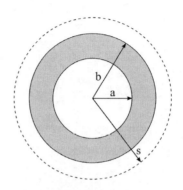

Again,

$$\oint_S \vec{B} \cdot d\vec{\ell} = B 2\pi s$$

with

$$I_{enc} = \int \vec{J} \cdot d\vec{a} = \int_a^b Js \, ds \int_0^{2\pi} d\phi = 2\pi \int_a^b ks \, s \, ds = 2\pi \frac{ks^3}{3}\Big|_a^b = \frac{2\pi k(b^3 - a^3)}{3}.$$

Therefore,

$$B = \frac{k(b^3 - a^3)}{3s}$$

with

$$\vec{B} = \frac{k(b^3 - a^3)}{3s}\hat{\phi}.$$

Problem 4.13. Find the vector potential d above a spinning disk of radius R, with angular velocity ω and carrying σ.

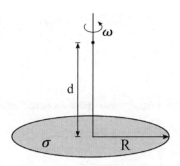

Solution We have

$$\vec{A} = \frac{\mu_0}{4\pi} \int \frac{\vec{K}}{\mathfrak{r}} da,$$

where

$$\vec{K} = \sigma \vec{v} = \sigma \omega r \hat{\phi}.$$

From

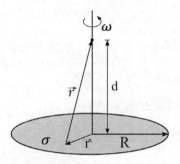

we also have

$$\mathfrak{r} = \sqrt{r^2 + d^2}$$

and

$$da = 2\pi r \, dr.$$

So

$$\vec{A} = \frac{\mu_0}{4\pi} \int_0^R \frac{\sigma \omega r 2\pi r \hat{\phi}}{\sqrt{r^2 + d^2}} dr = \frac{\mu_0 \sigma \omega \hat{\phi}}{2} \int_0^R \frac{r^2}{\sqrt{r^2 + d^2}} dr$$

$$= \frac{\mu_0 \sigma \omega}{4} \left[R\sqrt{R^2 + d^2} - d^2 \ln\left(R + \sqrt{R^2 + d^2}\right) + d^2 \ln d \right] \hat{\phi}$$

$$\vec{A} = \frac{\mu_0 \sigma \omega}{4} \left[R\sqrt{R^2 + d^2} + \ln\left(\frac{d}{R + \sqrt{R^2 + d^2}}\right) \right] \hat{\phi}.$$

Problem 4.14. What current density produces vector potential $\vec{A} = \sin\phi\, \hat{z}$?

Solution First, check $\nabla \cdot \vec{A} = 0$:

$$\nabla \cdot \vec{A} = \nabla \cdot (\sin\phi\, \hat{z}) = \frac{\partial}{\partial z}(\sin\phi) = 0.$$

Now, $\vec{B} = \nabla \times \vec{A}$ and $\nabla \times \vec{B} = \mu_0 \vec{J}$, so

$$\vec{B} = \nabla \times \vec{A} = \nabla \times (\sin\phi\, \hat{z}) = \frac{1}{s}\left[\frac{\partial}{\partial \phi}(\sin\phi)\right]\hat{s} = \frac{\cos\phi}{s}\hat{s}.$$

Also,

$$\mu_0 \vec{J} = \nabla \times \vec{B} = \nabla \times \left(\frac{\cos\phi}{s}\hat{s}\right) = \frac{1}{s}\left[-\frac{\partial}{\partial \phi}\left(\frac{\cos\phi}{s}\right)\right]\hat{z} = \frac{\sin\phi}{s^2}\hat{z}.$$

Therefore,

$$\vec{J} = \frac{\sin\phi}{\mu_0 s^2}\hat{z}.$$

This can be checked using the product rule

$$\nabla \times (\nabla \times \vec{A}) = \mu_0 \vec{J} = \nabla(\nabla \cdot \vec{A}) - \nabla^2 \vec{A}.$$

Since $\nabla \cdot \vec{A} = 0$, $\nabla^2 \vec{A} = \mu_0 \vec{J}$ where

$$-\nabla^2 \vec{A} = -\frac{1}{s}\frac{\partial^2}{\partial \phi^2}(\sin\phi)\hat{z} = -\frac{1}{s^2}(-\sin\phi)\hat{z} = \frac{\sin\phi}{s^2}\hat{z}.$$

So

$$\vec{J} = \frac{1}{\mu_0}\left(-\nabla^2 \vec{A}\right) = \frac{\sin\phi}{\mu_0 s^2}\hat{z}$$

as expected.

Problem 4.15. Find the vector potential inside and outside a wire of radius R that is carrying current density $\vec{J} = ks\hat{z}$, where k is a constant.

Solution We can find the field inside by

$$\oint_S \vec{B} \cdot d\vec{\ell} = \mu_0 I_{\text{enc}},$$

where

$$\oint_S \vec{B} \cdot d\vec{\ell} = B 2\pi s$$

and

$$I_{\text{enc}} = \int J \, da = \int_0^s ks' 2\pi s' \, ds' = \frac{2\pi k s^3}{3}.$$

So,

$$\oint_S \vec{B} \cdot d\vec{\ell} = B 2\pi s = \mu_0 \frac{2\pi k s^3}{3}$$

and

$$\vec{B} = \frac{\mu_0 k s^2}{3} \hat{\phi}.$$

We must have that $\vec{A}$ depends only on s and is in the direction of the current. So $\vec{A} = A(s)\hat{z}$ and $\nabla \times \vec{A} = \vec{B}$. Note

$$\nabla \times \vec{A} = -\frac{\partial A}{\partial s} \hat{\phi} = \vec{B} = \frac{\mu_0 k s^2}{3} \hat{\phi}.$$

Therefore,

$$dA = -\frac{\mu_0 k}{3} s^2$$

and

$$\vec{A} = \left(-\frac{\mu_0 k s^3}{9} + C \right) \hat{z}.$$

We will express this as

$$\vec{A} = -\frac{\mu_0 k}{9} \left(s^3 - \alpha^3 \right) \hat{z}.$$

Outside, our total current is

$$I_{\text{tot}} = I_{\text{enc}} = \int J \, da = \frac{2\pi k R^3}{3}.$$

From
$$\oint_S \vec{B}\cdot d\vec{\ell} = \mu_0 I_{\text{enc}}$$
we have
$$B 2\pi s = \mu_0 \frac{2\pi k R^3}{3}$$
and
$$\vec{B} = \frac{\mu_0 k R^3}{3s}\hat{\phi}.$$
Again,
$$dA = -B\,ds = -\frac{\mu_0 k R^3}{3s}ds$$
so
$$\vec{A} = \left(-\frac{\mu_0 k R^3}{3}\ln s + C\right)\hat{z}.$$
We will express this as
$$\vec{A} = -\frac{\mu_0 k R^3}{3}\left(\ln\frac{s}{\beta}\right)\hat{z}.$$
Since $\vec{A}$ is continuous at R,
$$-\frac{\mu_0 k}{9}(s^3 - a^3) = -\frac{\mu_0 k R^3}{3}\left(\ln\frac{s}{\beta}\right)$$
we have
$$R^3 - a^3 = 3R^3 \ln\frac{R}{\beta}$$
with
$$R^3\left(1 - 3\ln\frac{R}{\beta}\right) = a^3$$
and
$$1 - 3\ln\frac{R}{\beta} = \frac{a^3}{R^3}.$$

If $\alpha = \beta = R$,

$$1 - 3\ln\left(\frac{R}{R}\right) = \frac{R^3}{R^3} \to 1 = 1.$$

So

$$\vec{A} = \begin{cases} -\dfrac{\mu_0 k}{9}(s^3 - R^3)\hat{z} & s < R \\ -\dfrac{\mu_0 k R^3}{3}\ln\dfrac{s}{R}\hat{z} & s > R \end{cases}.$$

Problem 4.16. A disk of radius R is carrying surface charge $\sigma = kr$, where k is a constant, and spinning at angular velocity ω. Find the magnetic dipole moment and the field it produces.

Solution We have

$$\vec{K} = \sigma\vec{v} = \sigma\omega r\hat{\phi} = \omega k r^2 \hat{\phi}.$$

So

$$dI = \omega k r^2 dr \to I = \frac{\omega k r^3}{3}.$$

Therefore,

$$\vec{m} = \int I\, d\vec{a} = \frac{\omega k}{3}\int_0^R r^3 2\pi r\, dr\, \hat{z} = \frac{2\pi\omega k}{3}\int_0^R r^4\, dr$$

$$\vec{m} = \frac{2\pi\omega k R^5}{15}\hat{z}.$$

We have, in spherical coordinates,

$$\vec{B}_{\text{dip}} = \frac{\mu_0 m}{4\pi r^3}\left(2\cos\theta\,\hat{r} + \sin\theta\,\hat{\theta}\right),$$

which can be expressed in cylindrical coordinates by considering $r = \sqrt{s^2 + z^2}$, $\cos\theta = \frac{z}{r}$, $\sin\theta = \frac{s}{r}$, $\hat{\theta} = \frac{z}{r}\hat{s} - \frac{s}{r}\hat{z}$, and $\hat{r} = \frac{s}{r}\hat{s} + \frac{z}{r}\hat{z}$. Therefore,

$$\vec{B}_{\text{dip}} = \frac{\mu_0 m}{4\pi r^3}\left[2\left(\frac{z}{r}\right)\left(\frac{s}{r}\hat{s} + \frac{z}{r}\hat{z}\right) + \left(\frac{s}{r}\right)\left(\frac{z}{r}\hat{s} - \frac{s}{r}\hat{z}\right)\right]$$

$$= \frac{\mu_0 m}{4\pi r^5}\left(2zs\hat{s} + 2z^2\hat{z} + sz\hat{s} - s^2\hat{z}\right)$$

$$= \frac{\mu_0 m}{4\pi r^5}\Big[3zs\hat{s} + \big(2z^2 - s^2\big)\hat{z}\Big]$$

$$\vec{B}_{\text{dip}} = \frac{\mu_0 m}{4\pi} \frac{1}{\big(s^2 + z^2\big)^{5/2}}\Big[3zs\hat{s} + \big(2z^2 - s^2\big)\hat{z}\Big].$$

Substitution of m yields,

$$\vec{B}_{\text{dip}} = \frac{\mu_0 \omega k R^5}{30} \frac{1}{\big(s^2 + z^2\big)^{5/2}}\Big[3zs\hat{s} + \big(2z^2 - s^2\big)\hat{z}\Big].$$

Bibliography

Griffiths D J 1999 *Introduction to Electrodynamics* 3rd edn (Englewood Cliffs, NJ: Prentice Hall)
Griffiths D J 2013 *Introduction to Electrodynamics* 4th edn (New York: Pearson)
Halliday D, Resnick R and Walker J 2010 *Fundamentals of Physics* 9th edn (New York: Wiley)
Halliday D, Resnick R and Walker J 2013 *Fundamentals of Physics* 10th edn (New York: Wiley)

IOP Concise Physics

Electromagnetism
Problems and solutions
Carolina C Ilie and Zachariah S Schrecengost

Chapter 5

Electric fields in matter

Now we will address problems that deal with electric fields in matter, looking at problems involving dipole moments, media polarization, and electric displacement. Ideas developed in chapters 2 and 3 will be revisited and expanded upon in this chapter. Gauss's law is reformulated for electric displacement and various ways to calculate the energy of a configuration. Some of the techniques practiced in chapter 3 will be applied now, including the Laplace equation and Legendre polynomials.

5.1 Theory

5.1.1 Induced dipole moment of an atom in an electric field

Given an atom with polarizability α in an electric field $\vec{E}$, the induced dipole moment is

$$\vec{p} = \alpha \vec{E}.$$

5.1.2 Torque on a dipole due to an electric field

Given a dipole moment $\vec{p}$ in an electric field $\vec{E}$, the torque on the dipole is

$$\vec{N} = \vec{p} \times \vec{E}.$$

5.1.3 Force on a dipole

Given a dipole moment $\vec{p}$ in an electric field $\vec{E}$, the force on the dipole is

$$\vec{F} = (\vec{p} \cdot \nabla)\vec{E}.$$

5.1.4 Energy of a dipole in an electric field
Given a dipole moment $\vec{p}$ in an electric field $\vec{E}$, the energy of the dipole is
$$U = -\vec{p} \cdot \vec{E}.$$

5.1.5 Surface bound charge due to polarization $\vec{P}$
Given polarization $\vec{P}$ and normal vector $\hat{n}$, the surface bound charge is
$$\sigma_b = \vec{P} \cdot \hat{n}.$$

5.1.6 Volume bound charge due to polarization $\vec{P}$
Given polarization $\vec{P}$, the volume bound charge is
$$\rho_b = -\nabla \cdot \vec{P}.$$

5.1.7 Potential due to polarization $\vec{P}$
Given a volume $\mathcal{V}$, the potential due to polarization $\vec{P}(\vec{r})$ is
$$V(\vec{r}) = \frac{1}{4\pi\varepsilon_0} \int_\mathcal{V} \frac{\hat{r} \cdot \vec{P}(\vec{r}')}{r^2} d\tau'.$$

5.1.8. Electric displacement
Given polarization $\vec{P}$ and electric field $\vec{E}$, the electric displacement is given by
$$\vec{D} = \varepsilon_0 \vec{E} + \vec{P}.$$

5.1.9 Gauss's law for electric displacement
Considering electric displacement $\vec{D}$ and free charge density ρ_f, Gauss's law can be written in differential form as
$$\nabla \cdot \vec{D} = \rho_f$$
and in integral form as
$$\oint_S \vec{D} \cdot d\vec{a} = q_{f_{enc}},$$
where $q_{f_{enc}}$ is the total free charge enclosed in the volume.

5.1.10 Linear dielectrics
Given a medium with electric susceptibility χ_e, the polarization is given by
$$\vec{P} = \varepsilon_0 \chi_e \vec{E},$$

where $\vec{E}$ is the total electric field. The electric displacement is now

$$\vec{D} = \varepsilon_0 \vec{E} + \vec{P} = \left(\varepsilon_0 + \varepsilon_0 \chi_e\right)\vec{E} = \varepsilon_0\left(1 + \chi_e\right)\vec{E} = \varepsilon_0 \varepsilon_r \vec{E} = \varepsilon \vec{E},$$

where ε is the permittivity of the material and ε_r is the relative permittivity of the material. Also, the boundary conditions are now

$$\varepsilon_{\text{above}} E^{\perp}_{\text{above}} - \varepsilon_{\text{below}} E^{\perp}_{\text{below}} = \sigma_f$$

And

$$\varepsilon_{\text{above}}\frac{\partial V_{\text{above}}}{\partial n} - \varepsilon_{\text{below}}\frac{\partial V_{\text{below}}}{\partial n} = -\sigma_f$$

while we still maintain

$$V_{\text{above}} = V_{\text{below}}.$$

5.1.11 Energy in a dielectric system

Given electric field $\vec{E}$ and electric displacement $\vec{D}$, the energy in a dielectric system is

$$W = \frac{\varepsilon_0}{2}\int \varepsilon_r E^2 d\tau = \frac{1}{2}\int \vec{D}\cdot\vec{E}\, d\tau.$$

5.2 Problems and solutions

Problem 5.1. Given $\vec{p}_1$ and $\vec{p}_2$ below, find where to place point charge q such that there is no net torque on $\vec{p}_2$. Assume the center of $\vec{p}_1$ is the origin and express your answer in spherical coordinates.

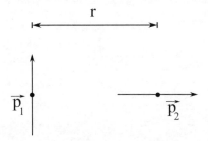

Solution The field at $\vec{p}_2$ is given by

$$\vec{E}_{\text{dip}} = \frac{p_1}{4\pi\varepsilon_0 r^3}\hat{\theta} = -\frac{p_1}{4\pi\varepsilon_0 r^3}\hat{z}.$$

Since the field is straight down, we must place the point charge q below it to cancel the field (thus resulting in zero torque on $\vec{p}_2$). We must place q at a distance d from the dipole so that

$$E_{\text{dip}} + E_q = -\frac{p_1}{4\pi\varepsilon_0 r^3} + \frac{q}{4\pi\varepsilon_0 d^2} = 0.$$

Solving for d^2 yields

$$d^2 = \frac{qr^3}{p_1}.$$

Therefore we have

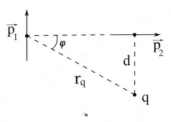

with

$$r_q^2 = \frac{qr^3}{p_1} + r^2 = r^2\left(\frac{qr}{p_1} + 1\right)$$

so

$$r_q = r\sqrt{\frac{qr}{p_1} + 1}.$$

From

$$\cos\varphi = \frac{r}{r_q}$$

we have

$$\varphi = \cos^{-1}\left(\frac{1}{\sqrt{\frac{qr}{p_1} + 1}}\right).$$

So the spherical coordinates of q are

$$(r, \theta) = \left(r\sqrt{\frac{qr}{p_1} + 1}, \frac{\pi}{2} + \cos^{-1}\left(\frac{1}{\sqrt{\frac{qr}{p_1} + 1}}\right) \right).$$

Problem 5.2. Consider a neutral atom, with polarizability α, located z above a disk of radius R carrying surface charge σ. Find the force of attraction between the atom and the plate.

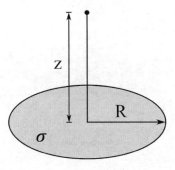

Solution The field at z is given by

$$\vec{E} = \frac{1}{4\pi\varepsilon_0} \int \frac{\sigma}{r^2} \hat{r} \, da = \frac{\sigma}{4\pi\varepsilon_0} \int_0^R \frac{2\pi r z}{(z^2 + r^2)^{3/2}} \hat{z} \, dr$$

$$\vec{E} = \frac{z\sigma}{2\varepsilon_0} \left(\frac{1}{z} - \frac{1}{\sqrt{R^2 + z^2}} \right) \hat{z}.$$

This induces a dipole

$$\vec{p} = \alpha \vec{E} = \frac{z\sigma\alpha}{2\varepsilon_0} \left(\frac{1}{z} - \frac{1}{\sqrt{R^2 + z^2}} \right) \hat{z}.$$

The electric field due to the dipole is given by

$$\vec{E}_{\text{dip}} = \frac{p}{4\pi\varepsilon_0 r^3} \left(2\cos\theta \, \hat{r} + \sin\theta \, \hat{\theta} \right)$$

$$= \frac{z\sigma\alpha}{8\pi\varepsilon_0^2 r^3} \left(\frac{1}{z} - \frac{1}{\sqrt{R^2 + z^2}} \right) \left(2\cos\theta \, \hat{r} + \sin\theta \, \hat{\theta} \right).$$

The force on a piece of charge dq is given by
$$d\vec{F} = \vec{E} dq$$
so
$$\vec{F} = \int d\vec{F} = \int \vec{E} dq = \int \vec{E} \sigma \, dA = \int \vec{E} \sigma \, 2\pi \ell \, d\ell.$$

Consider the following

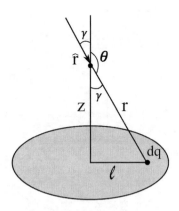

with side view

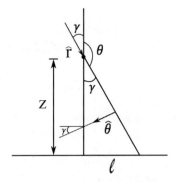

In our expression fix alignment for $\vec{E}$, we have the term $(2\cos\theta \, \hat{r} + \sin\theta \, \hat{\theta})$. Due to symmetry, we only have a $\hat{z}$-component of the force. Therefore,

$$\hat{r} \to -\cos\gamma \, \hat{z} = -\frac{z}{r}\hat{z}$$

and
$$\hat{\theta} = -\sin\gamma\,\hat{z} = -\frac{\ell}{r}\hat{z},$$
where
$$r = \sqrt{z^2 + \ell^2}.$$

Also, we have $\theta = \pi - \gamma$. So
$$\cos\theta = \cos(\pi - \gamma) = \cos(\pi)\cos(-\gamma) + \sin(\pi)\sin(-\gamma) = -\cos\gamma = -\frac{z}{r}$$
and
$$\sin\theta = \sin(\pi - \gamma) = \sin(\pi)\cos(-\gamma) + \cos(\pi)\sin(-\gamma) = -\sin(-\gamma) = \sin\gamma = \frac{\ell}{r}.$$

Therefore, we have
$$2\cos\theta\,\hat{r} + \sin\theta\,\hat{\theta} = 2\left(-\frac{z}{r}\right)\left(-\frac{z}{r}\right)\hat{z} + \left(\frac{\ell}{r}\right)\left(-\frac{\ell}{r}\right)\hat{z} = -\frac{1}{r^2}\left(-2z^2 + \ell^2\right)\hat{z}.$$

Our force becomes
$$\vec{F} = \frac{2\pi z \sigma^2 \alpha}{8\pi\varepsilon_0^2}\left(\frac{1}{z} - \frac{1}{\sqrt{R^2 + z^2}}\right)\int_0^R -\frac{(-2z^2 + \ell^2)\ell}{(z^2 + \ell^2)^{5/2}}\hat{z}\,d\ell$$

$$\vec{F} = \frac{z\sigma^2\alpha}{4\varepsilon_0^2}\left(\frac{1}{z} - \frac{1}{\sqrt{R^2 + z^2}}\right)\left[\frac{R^2}{(R^2 + z^2)^{3/2}}\right]\hat{z} = \frac{\sigma^2\alpha R^2\left(\sqrt{R^2 + z^2} - z\right)}{4\varepsilon_0^2(R^2 + z)^2}\hat{z}.$$

So at $z = d$, the force of attraction is
$$F = \frac{\sigma^2\alpha R^2\left(\sqrt{R^2 + d^2} - d\right)}{4\varepsilon_0^2(R^2 + d)^2}.$$

We can verify this using
$$\vec{F} = (\vec{p}\cdot\nabla)\vec{E}$$

Note
$$\vec{p}\cdot\nabla = \frac{\sigma\alpha}{2\varepsilon_0}\left(1 - \frac{z}{\sqrt{R^2 + z^2}}\right)\frac{\partial}{\partial z}$$

and

$$\vec{F} = (\vec{p} \cdot \nabla)\vec{E} = \frac{\sigma\alpha}{2\varepsilon_0}\left(1 - \frac{z}{\sqrt{R^2 + z^2}}\right)\frac{\partial}{\partial z}\left[\frac{\sigma}{2\varepsilon_0}\left(1 - \frac{z}{\sqrt{R^2 + z^2}}\right)\hat{z}\right]$$

$$\vec{F} = -\frac{z\sigma^2\alpha}{4\varepsilon_0^2}\left(\frac{1}{z} - \frac{1}{\sqrt{R^2 + z^2}}\right)\left[\frac{R^2}{(R^2 + z^2)^{3/2}}\right]\hat{z} = -\frac{\sigma^2\alpha R^2\left(\sqrt{R^2 + z^2} - z\right)}{4\varepsilon_0^2\left(R^2 + z\right)^2}\hat{z},$$

which is equal in magnitude and opposite in direction to what was found above. Why is this? In the first method, we calculated the force on the plate *from* the atom. So a positive force is 'attractive'. In the second method, we are finding the force on the atom from the plate. A negative force at the atom 'attracts' it to the plate.

Problem 5.3. Consider $\vec{p}_1$ and $\vec{p}_2$ below. Find the force of $\vec{p}_2$ on $\vec{p}_1$ and verify using the energy of the configuration.

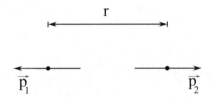

Solution The electric field due to a dipole is

$$\vec{E}_{\text{dip}}(r, \theta) = \frac{p}{4\pi\varepsilon_0 r^3}\left(2\cos\theta\,\hat{r} + \sin\theta\,\hat{\theta}\right).$$

Here, $\theta = \pi$, so the field at $\vec{p}_1$ due to $\vec{p}_2$ is

$$\vec{E}_{\text{dip}} = \frac{-2p_2}{4\pi\varepsilon_0 r^3}\hat{r} = \frac{-2p_2}{4\pi\varepsilon_0 y^3}\hat{y}.$$

The force is given by

$$\vec{F} = (\vec{p} \cdot \nabla)\vec{E}$$

where $\vec{p} = \vec{p}_1 = -p_1\hat{y}$. So

$$\vec{p} \cdot \nabla = -p_1\frac{\partial}{\partial y}.$$

Therefore,
$$\vec{F} = (\vec{p} \cdot \nabla)\vec{E} = -p_1 \frac{\partial}{\partial y}\left(\frac{-2p_2}{4\pi\varepsilon_0 y^3}\hat{y}\right) = \frac{2p_1 p_2}{4\pi\varepsilon_0}\frac{\partial}{\partial y}\left(y^{-3}\right)\hat{y}$$

$$\vec{F} = -\frac{3p_1 p_2}{2\pi\varepsilon_0 y^4}\hat{y}.$$

The energy stored in this configuration is given by
$$U = -\vec{p}_1 \cdot \vec{E} = -\left[(-p_1 \hat{y}) \cdot \left(\frac{-2p_2}{4\pi\varepsilon_0 y^3}\hat{y}\right)\right] = -(-p_1)\left(\frac{-2p_2}{4\pi\varepsilon_0 y^3}\right) = -\frac{2p_1 p_2}{4\pi\varepsilon_0 y^3}.$$

From this, the force is given by
$$\vec{F} = -\nabla U = -\nabla\left(\frac{-2p_1 p_2}{4\pi\varepsilon_0 y^3}\right) = \frac{2p_1 p_2}{4\pi\varepsilon_0}\frac{\partial}{\partial y}\left(y^{-3}\right)\hat{y}$$

$$\vec{F} = -\frac{3p_1 p_2}{2\pi\varepsilon_0 y^4}\hat{y}$$

as expected.

Problem 5.4. Consider the two dipoles depicted below. Find the angle γ that maximizes and minimizes the magnitude of the torque on $\vec{p}_2$ due to $\vec{p}_1$.

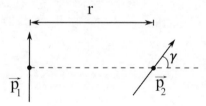

Solution The electric field due to a dipole is
$$\vec{E}_{\text{dip}}(r, \theta) = \frac{p}{4\pi\varepsilon_0 r^3}\left(2\cos\theta\,\hat{r} + \sin\theta\,\hat{\theta}\right).$$

Here, $\theta = \frac{\pi}{2}$, so the field at $\vec{p}_2$ due to $\vec{p}_1$ is
$$\vec{E}_{\text{dip}} = \frac{p_1}{4\pi\varepsilon_0 r^3}\hat{\theta},$$

which points down. The torque is given by
$$\vec{N} = \vec{p}_2 \times \vec{E}_{\text{dip}}.$$

We can express $\vec{E}_{dip}$ and $\vec{p}_2$ by

$$\vec{E}_{dip} = -\frac{p_1}{4\pi\varepsilon_0 y^3}\hat{z}$$

and

$$\vec{p}_2 = p_2 \cos\gamma\, \hat{y} + p_2 \sin\gamma\, \hat{z}.$$

So

$$\vec{N} = \vec{p}_2 \times \vec{E}_{dip} = \begin{vmatrix} \hat{x} & \hat{y} & \hat{z} \\ 0 & p_2\cos\gamma & p_2\sin\gamma \\ 0 & 0 & -\frac{p_1}{4\pi\varepsilon_0 y^3} \end{vmatrix} = \frac{-p_1 p_2 \cos\gamma}{4\pi\varepsilon_0 y^3}\hat{x}.$$

We can see that $|\vec{N}|$ is maximum when $\gamma = 0$ and $\gamma = \pi$

$$|\vec{N}| = \frac{p_1 p_2}{4\pi\varepsilon_0 y^3}$$

and minimum when $\gamma = \frac{\pi}{2}$ and $\gamma = \frac{3\pi}{2}$

$$|\vec{N}| = 0.$$

Note the effect of aligning the dipole parallel to the field, $|\vec{p} \times \vec{E}| = 0$, and aligning the dipole perpendicular to the field $|\vec{p} \times \vec{E}| = pE$.

Problem 5.5. A sphere of radius R carries polarization $\vec{P} = \frac{k}{r}\hat{r}$, where k is a constant, from $r = a$ to $r = R$. Find the electric field in all regions.

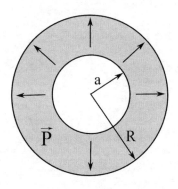

Solution The bound charges are given by

$$\sigma_b(a) = \vec{P} \cdot \hat{n} = \frac{k}{a}\hat{r} \cdot (-\hat{r}) = -\frac{k}{a}$$

$$\sigma_b(R) = \vec{P} \cdot \hat{n} = \frac{k}{R}\hat{r} \cdot \hat{r} = \frac{k}{R}$$

$$\rho_b = -\nabla \cdot \vec{P} = -\frac{1}{r^2}\frac{\partial}{\partial r}\left(r^2\frac{k}{r}\right) = -\frac{k}{r^2}.$$

When $r < a$, $q_{enc} = 0$, so $\vec{E} = 0$. When $a \leqslant r \leqslant R$, we have

$$\oint_S \vec{E} \cdot d\vec{a} = \frac{q_{enc}}{\varepsilon_0}$$

with

$$q_{enc} = -\frac{k}{a}(4\pi a^2) + 4\pi \int_a^r -\frac{k}{(r')^2}(r')^2 dr' = -4\pi kr.$$

So

$$\oint_S \vec{E} \cdot d\vec{a} = E4\pi r^2 = -\frac{4\pi kr}{\varepsilon_0}$$

and

$$\vec{E} = -\frac{k}{\varepsilon_0}\frac{1}{r}\hat{r} = -\frac{k}{\varepsilon_0}\frac{\vec{r}}{r^2}.$$

When $r > R$, we have

$$q_{enc} = -\frac{k}{a}(4\pi a^2) + 4\pi \int_a^R -\frac{k}{(r')^2}(r')^2 \, dr' + \frac{k}{R}(4\pi R^2) = 0$$

so $\vec{E} = 0$.

Problem 5.6. Consider a very long cylinder of radius R hollowed out to a radius a and carrying a uniform, radial polarization $\vec{P}$ and charge density $\rho = ks$, where k is a constant. Find the electric field in all three regions ($\vec{P}$ from a to R, ρ from 0 to R).

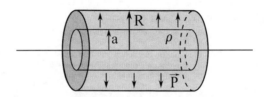

Solution For $s < a$, all we have is charge, so
$$\oint_S \vec{E} \cdot d\vec{a} = \frac{q_{enc}}{\varepsilon_0},$$
where
$$q_{enc} = 2\pi\ell \int_0^s ks's' ds' = \frac{2\pi\ell k s^3}{3}.$$
So
$$\oint_S \vec{E} \cdot d\vec{a} = E 2\pi s\ell = \frac{2\pi\ell k s^3}{3\varepsilon_0}$$
and
$$\vec{E} = \frac{s^2 k}{3\varepsilon_0}\hat{s}.$$

For $a \leqslant s \leqslant R$, we have bound charge
$$\sigma_b(a) = \vec{P} \cdot \hat{n} = P\hat{s} \cdot (-\hat{s}) = -P$$
$$\rho_b = -\nabla \cdot \vec{P} = -\frac{1}{s}\frac{\partial}{\partial s}(sP) = -\frac{P}{s}.$$
So,
$$q_{enc} = -P 2\pi a\ell + \frac{2\pi\ell k s^3}{3} + 2\pi\ell \int_a^s -\frac{P}{s'} s' \, ds' = 2\pi\ell\left(\frac{ks^3}{3} - Pa - Ps + Pa\right)$$
$$q_{enc} = 2\pi\ell s\left(\frac{ks^2}{3} - P\right).$$
Therefore,
$$\oint_S \vec{E} \cdot d\vec{a} = E 2\pi s\ell = \frac{2\pi\ell s}{\varepsilon_0}\left(\frac{ks^2}{3} - P\right)$$

So
$$\vec{E} = \frac{1}{3\varepsilon_0}(ks^2 - 3P)\hat{s}.$$

For $s > R$, we have
$$\sigma_b(R) = \vec{P} \cdot \hat{n} = P\hat{s} \cdot \hat{s} = P.$$

So,
$$q_{enc} = -P2\pi a \ell + \frac{2\pi \ell k R^3}{3} + 2\pi \ell \int_a^R -\frac{P}{s'}s'\,ds' + P2\pi R \ell$$

$$= 2\pi \ell \left(\frac{kR^3}{3} - Pa - PR + Pa + PR\right)$$

$$q_{enc} = \frac{2\pi \ell k R^3}{3}.$$

Therefore,
$$\oint_S \vec{E} \cdot d\vec{a} = E 2\pi s \ell = \frac{2\pi \ell k R^3}{3\varepsilon_0}$$

and
$$\vec{E} = \frac{kR^3}{3s\varepsilon_0}\hat{s}.$$

Problem 5.7. Consider a cylinder of radius R and length L, carrying polarization $\vec{P} = P\hat{z}$. Find the potential d from the cylinder.

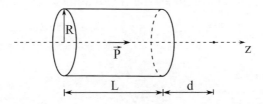

Solution The potential is
$$V(\vec{r}) = \frac{1}{4\pi\varepsilon_0} \int_v \frac{\hat{r} \cdot \vec{P}}{r^2} d\tau.$$

We can see from

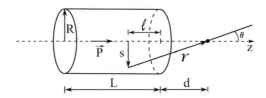

that
$$r = \sqrt{(\ell + d)^2 + s^2}.$$

If $z = 0$ is the left side of the cylinder, then ℓ goes from L to 0. So
$$r = \sqrt{(L - z + d)^2 + s^2}.$$

Also,
$$\hat{r} = \cos\theta\,\hat{z} = \frac{\ell + d}{r}\hat{z} = \frac{L - z + d}{r}\hat{z}.$$

Since $d\tau = s\,ds\,d\phi\,dz$, we have
$$V = \frac{1}{4\pi\varepsilon_0} \int_0^L \int_0^R \int_0^{2\pi} \frac{P(L - z + d)s}{\left[(L - z + d)^2 + s^2\right]^{3/2}} d\phi\,ds\,dz.$$

Using $u = L - z + d$, we have $du = -dz$. Evaluating u at the endpoints yields
$$u(z = 0) = L + d$$

And
$$u(z = L) = d.$$

So
$$V = \frac{P}{2\varepsilon_0} \int_d^{L+d} \int_0^R \frac{us}{(u^2 + s^2)^{3/2}} ds\,du = \frac{P}{2\varepsilon_0} \int_d^{L+d} \left(1 - \frac{u}{\sqrt{R^2 + u^2}}\right) du.$$

Using $x = R^2 + u^2$, we have $dx = 2u\, du$ and

$$V = \frac{P}{2\varepsilon_0}\left(u\Big|_d^{L+d} - \int_{R^2+d^2}^{R^2+(L+d)^2} \frac{1}{2}x^{-\frac{1}{2}}dx\right)$$

$$= \frac{P}{2\varepsilon_0}\left[L + d - d - \sqrt{R^2 + (L+d)^2} + \sqrt{R^2 + d^2}\right]$$

$$V = \frac{P}{2\varepsilon_0}\left[L - \sqrt{R^2 + (L+d)^2} + \sqrt{R^2 + d^2}\right].$$

Problem 5.8. Consider a sphere of radius R carrying polarization $\vec{P}(\vec{r}) = kr^n\hat{r}$ where n is an integer and k is a constant. Find the charge density required to cancel the polarization.

Solution The bound charge is given by

$$\rho_b = -\nabla \cdot \vec{P} = -\frac{1}{r^2}\frac{\partial}{\partial r}\left(r^2 kr^n\right)$$

$$= -\frac{k}{r^2}\frac{\partial}{\partial r}\left(r^{n+2}\right) = -\frac{k(n+2)}{r^2}r^{n+1} = -k(n+2)r^{n-1}.$$

Therefore, a charge density

$$\rho = k(n+2)r^{n-1}$$

will cancel the bound charge produced by $\vec{P}(\vec{r}) = kr^n\hat{r}$.

Problem 5.9. A spherical shell of radius R with surface charge density σ is surrounded up to radius a by an LIH dielectric material of susceptibility χ_e. Find the electric displacement and the electric field.

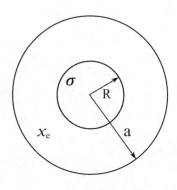

Solution Gauss's law for electric displacement is given by

$$\oint_S \vec{D} \cdot d\vec{a} = q_{f_{enc}}.$$

For $r < R$, we have

$$q_{f_{enc}} = 0.$$

Therefore,

$$D = 0.$$

For $r > R$, the left-hand side of Gauss's law is given by

$$\oint_S \vec{D} \cdot d\vec{a} = \oint_S D \, da = D \oint_S da = D 4\pi r^2.$$

Now, the enclosed free charge is

$$q_{f_{enc}} = \sigma 4\pi R^2.$$

So

$$D 4\pi r^2 = \sigma 4\pi R^2$$

and

$$D = \frac{\sigma R^2}{r^2} \rightarrow \vec{D} = \frac{\sigma R^2}{r^2} \hat{r}.$$

Let us consider the electric field. For $r < R$, we have

$$\vec{E} = 0$$

Since

$$\vec{D} = \varepsilon \vec{E},$$

for $R < r < a$, the electric displacement is

$$D = \frac{\sigma R^2}{r^2},$$

so the electric field is

$$\vec{E} = \frac{\vec{D}}{\varepsilon} = \frac{\vec{D}}{\varepsilon_0 \varepsilon_r} = \frac{\vec{D}}{\varepsilon_0 (1 + \chi_e)} = \frac{\sigma R^2}{\varepsilon_0 (1 + \chi_e) r^2} \hat{r}.$$

Finally, for $r > a$, the displacement is

$$D = \frac{\sigma R^2}{r^2}$$

so

$$\vec{E} = \frac{\vec{D}}{\varepsilon_0} = \frac{\sigma R^2}{\varepsilon_0 r^2}\hat{r}.$$

Problem 5.10. For the previous problem, calculate the electric potential everywhere, relative to infinity.

Solution The electric field in the three regions is given by

$$E = \begin{cases} 0 & r < R \\ \dfrac{\sigma R^2}{\varepsilon_0(1+\chi_e)r^2} & R < r < a \\ \dfrac{\sigma R^2}{\varepsilon_0 r^2} & r > a \end{cases}.$$

For $r > a$,

$$V = -\int_\infty^r \vec{E}\cdot d\vec{\ell} = -\int_\infty^r \frac{\sigma R^2}{\varepsilon_0 r'^2}dr' = \left.\frac{\sigma R^2}{\varepsilon_0 r'}\right|_\infty^r = \frac{\sigma R^2}{\varepsilon_0 r}.$$

For $R < r < a$

$$V = -\int \vec{E}\cdot d\vec{\ell} = -\int_\infty^a \frac{\sigma R^2}{\varepsilon_0 r^2}dr - \int_a^r \frac{\sigma R^2}{\varepsilon_0(1+\chi_e)r'^2}dr'$$

$$= \frac{\sigma R^2}{\varepsilon_0 a} + \frac{\sigma R^2}{\varepsilon_0(1+\chi_e)}\left(\frac{1}{r} - \frac{1}{a}\right).$$

For $r < R$

$$V = -\int_\infty^a \frac{\sigma R^2}{\varepsilon_0 r^2}dr - \int_a^R \frac{\sigma R^2}{\varepsilon_0(1+\chi_e)r^2}dr - \int_R^0 0\,dr$$

$$= \frac{\sigma R^2}{\varepsilon_0 a} + \frac{\sigma R^2}{\varepsilon_0(1+\chi_e)}\left(\frac{1}{R} - \frac{1}{a}\right) = \text{const}.$$

Problem 5.11. A long cylinder of radius a carries a charge density that is proportional to the distance from the axis, $\rho = ks$, k constant. The cylinder is surrounded by rubber insulation out to a radius R. Find the electric displacement.

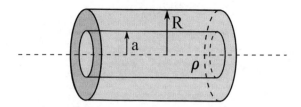

Solution For $s > a$, we have

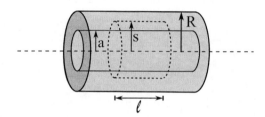

Gauss's law states

$$\oint_S \vec{D} \cdot d\vec{a} = q_{f_{enc}},$$

where the left-hand side is given by

$$\oint_S \vec{D} \cdot d\vec{a} = \oint_S D \, da = D \oint_S da = D2\pi s \ell$$

and the enclosed free charge is

$$q_{f_{enc}} = \int_V \rho \, d\tau = k \int_0^a s^2 ds \int_0^{2\pi} d\phi \int_0^\ell dz = \frac{2\pi k a^3 \ell}{3}.$$

So

$$D2\pi s \ell = \frac{2\pi k a^3 \ell}{3}$$

And

$$D = \frac{ka^3}{3s} \rightarrow \vec{D} = \frac{ka^3}{3s}\hat{s}.$$

From this, $\vec{E} = \frac{\vec{D}}{\varepsilon_0}$ for $s > R$; if we knew $\vec{P}$, we could find the electric field as well.

For $s < a$,

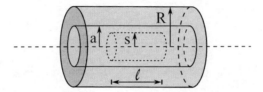

Now, the enclosed free charge is

$$q_{f_{enc}} = \int_V \rho \, d\tau = k \int_0^s s's' ds' \int_0^{2\pi} d\phi \int_0^{\ell} dz = \frac{2\pi k s^3 \ell}{3}.$$

Therefore,

$$D 2\pi s \ell = \frac{2\pi k s^3 \ell}{3}$$

and

$$D = \frac{ks^2}{3} \to \vec{D} = \frac{ks^2}{3} \hat{s}.$$

The electric field can be easily obtained,

$$\vec{E} = \frac{\vec{D}}{\varepsilon_0} = \frac{ks^2}{3\varepsilon_0} \hat{s}.$$

Problem 5.12. A sphere of radius R carries a polarization $\vec{P} = kr^2 \hat{r}$, k constant:
(a) Calculate the bound charges σ_b and ρ_b.
(b) Find the electric field inside and outside the sphere by using the bound charges and Gauss's law for $\vec{E}$.
(c) Calculate the field by using Gauss's law for $\vec{D}$.

Solutions
(a) Calculate the bound charges σ_b and ρ_b.
The bound charge density is given by

$$\rho_b = -\nabla \cdot \vec{P} = -\left[\frac{1}{r^2}\frac{\partial}{\partial r}(r^2 kr^2)\right] = -4kr$$

and the surface charge density is given by

$$\sigma_b = \vec{P} \cdot \hat{n} = kr^2 \hat{r} \cdot \hat{n}|_{r=R} = kR^2 = \text{const}.$$

(b) Find the electric field inside and outside the sphere by using the bound charges and Gauss's law for $\vec{E}$.
What is the total bound charge (what do you expect it to be)?

$$q_b = \sigma_b(\text{area}) + \int_V \rho_b \, d\tau = kR^2 4\pi R^2 + \int_0^R (-4kr) r^2 \, dr \int_0^{2\pi} d\phi \int_0^{\pi} \sin\theta \, d\theta$$

$$= 4\pi k R^4 - 4\pi 4k \frac{r^4}{4} \bigg|_0^R = 4\pi k R^4 - 4\pi k R^4 = 0.$$

For $r < R$, we have Gauss's law

$$\oint_S \vec{E} \cdot d\vec{a} = \frac{q_{\text{enc}}}{\varepsilon_0}$$

with the left-hand side given by

$$\oint_S \vec{E} \cdot d\vec{a} = E 4\pi r^2$$

and the enclosed charge

$$q_{\text{enc}} = \int_V \rho_b \, d\tau = \int_0^r (-4kr') r'^2 dr' \int_0^{2\pi} d\phi \int_0^{\pi} \sin\theta \, d\theta$$

$$= -4\pi 4k \frac{r'^4}{4} \bigg|_0^r = -4\pi k r^4.$$

Therefore,

$$E 4\pi r^2 = \frac{-4\pi k r^4}{\varepsilon_0}$$

with

$$E = \frac{kr^2}{\varepsilon_0} \to \vec{E} = \frac{kr^2}{\varepsilon_0} \hat{r}.$$

For $r > R$, we found $q_{\text{enc}} = 0$ so $\vec{E} = 0$.

(c) Calculate the field by using Gauss's law for $\vec{D}$.
To find the electric displacement, we consider Gauss's law for dielectrics

$$\oint_S \vec{D} \cdot d\vec{a} = q_{f_{\text{enc}}}.$$

However, we have no free charge, $q_{f_{enc}} = 0$, and $\vec{D} = 0$. Therefore, from

$$\vec{D} = \varepsilon_0 \vec{E} + \vec{P}$$

we have

$$\vec{E} = -\frac{\vec{P}}{\varepsilon_0}.$$

So

$$\vec{E} = -\frac{kr^2 \hat{r}}{\varepsilon_0} \quad (r < R)$$

and

$$\vec{E} = 0 \quad (r > R)$$

as expected.

Problem 5.13. A spherical conductor of radius R carries a surface charge density σ. The sphere is surrounded by a linear homogenous dielectric of susceptibility χ_e. Calculate the energy of this configuration.

Solution The energy is given by

$$W = \frac{\varepsilon_0}{2} \int \varepsilon_r E^2 d\tau = \frac{1}{2} \int \vec{D} \cdot \vec{E} d\tau.$$

It is very easy to obtain $\vec{D}$ and $\vec{E}$. For $r < R$, we have Gauss's law

$$\oint_S \vec{E} \cdot d\vec{a} = \frac{q_{enc}}{\varepsilon_0}$$

with $q_{enc} = 0$ so $\vec{E} = 0$. Also,

$$\oint_S \vec{D} \cdot d\vec{a} = q_{f_{enc}} = 0$$

So

$$\vec{D} = 0.$$

For $r > R$, the total enclosed charge is simply

$$q_{f_{enc}} = \sigma 4\pi R^2$$

and

$$\oint_S \vec{D} \cdot d\vec{a} = D 4\pi r^2.$$

So
$$D 4\pi r^2 = \sigma 4\pi R^2$$

With
$$D = \frac{\sigma R^2}{r^2}.$$

For $R < r < a$, the polarization is given by
$$\vec{P} = \varepsilon_0 \chi_e \vec{E}$$

so the electric displacement is
$$\vec{D} = \varepsilon_0 \vec{E} + \vec{P} = \varepsilon \vec{E}.$$

Solving for the electric field, we have
$$\vec{E} = \frac{\vec{D}}{\varepsilon} = \frac{\sigma R^2}{\varepsilon r^2} = \frac{\sigma R^2}{\varepsilon_0 (1 + \chi_e) r^2}.$$

For $r > a$, the field is just
$$E = \frac{\sigma R^2}{\varepsilon_0 r^2}.$$

Therefore, our electric displacements are
$$\vec{D} = \begin{cases} 0 & r < R \\ \dfrac{\sigma R^2}{r^2} & r > R \end{cases}$$

and our electric fields are
$$\vec{E} = \begin{cases} 0 & r < R \\ \dfrac{\sigma R^2}{\varepsilon_0 (1 + \chi_e) r^2} & R < r < a \\ \dfrac{\sigma R^2}{\varepsilon_0 r^2} & r > a \end{cases}.$$

Returning to the energy, we have
$$W = \frac{1}{2} \int \vec{D} \cdot \vec{E} \, d\tau = \frac{4\pi}{2} \int_R^a \frac{\sigma^2 R^4}{\varepsilon r^4} r^2 dr + \frac{4\pi}{2} \int_a^\infty \frac{\sigma^2 R^4}{\varepsilon_0 r^4} r^2 dr,$$

where we used
$$\int d\tau = 4\pi \int r^2 dr.$$

Therefore,
$$W = \frac{2\pi\sigma^2 R^4}{\varepsilon_0}\left[\frac{1}{\varepsilon_r}\left(\frac{1}{R}-\frac{1}{a}\right)+\frac{1}{a}\right] = \frac{2\pi\sigma^2 R^4}{\varepsilon_0 \varepsilon_r}\left(\frac{1}{R}+\frac{\chi_e}{a}\right) = \frac{2\pi\sigma^2 R^4}{\varepsilon_0(1+\chi_e)}\left(\frac{1}{R}+\frac{\chi_e}{a}\right).$$

Problem 5.14. A sphere of radius R, made of linear homogeneous dielectric material, is brought into a uniform electric field of magnitude $\vec{E}_0$. Using the Laplace equation and Legendre polynomials, find the electric field inside the sphere.

Solution In spherical coordinates, Laplace's equation is given by
$$\frac{1}{r^2}\frac{\partial}{\partial r}\left(r^2 \frac{\partial V}{\partial r}\right) + \frac{1}{r^2 \sin\theta}\frac{\partial}{\partial \theta}\left(\sin\theta \frac{\partial V}{\partial \theta}\right) + \frac{1}{r^2(\sin\theta)^2}\frac{\partial^2 V}{\partial \phi^2} = 0.$$

We have azimuthal symmetry, therefore the potential is ϕ independent, so
$$\frac{1}{r^2}\frac{\partial}{\partial r}\left(r^2 \frac{\partial V}{\partial r}\right) + \frac{1}{r^2 \sin\theta}\frac{\partial}{\partial \theta}\left(\sin\theta \frac{\partial V}{\partial \theta}\right) = 0.$$

We have outlined the solutions to this in chapter 3, and found the general solution to be given by
$$V(r,\theta) = \sum_{l=0}^{\infty}\left(A_l r^l + \frac{B_l}{r^{l+1}}\right)P_l(\cos\theta).$$

Now we can look at the boundary conditions for this particular problem. We need the electric potential to satisfy:
1) $V_{in} = V_{out}$ at $r = R$.
2) $\varepsilon_{above}E^{\perp}_{above} - \varepsilon_{below}E^{\perp}_{below} = \sigma_{free}$.
3) At large distance from the sphere: $r \gg R$, the potential must be $V_{out} = -E_0 r \cos\theta$.

Since the free surface charge density is zero, and by using the relationship between the electric field and the electric potential, the second condition becomes
$$\varepsilon \frac{\partial V_{in}}{\partial r} = \varepsilon_0 \frac{\partial V_{out}}{\partial r}$$

at $r = R$.

Now with clear boundary conditions and the general solution for the potential, we can write the potential inside the sphere and the potential outside the sphere. Looking at our general solution, we require $B_l = 0$ for $r < R$; otherwise $V \to \infty$ as $r \to 0$. Similarly, we require $A_l = 0$ for $r > R$; otherwise $V \to \infty$ as $r \to \infty$. For the potential outside the sphere we want to make sure we cover the third boundary condition, and this is why we will have two terms.

Inside the sphere, we have

$$V_{in}(r, \theta) = \sum_{l=0}^{\infty} A_l r^l P_l(\cos \theta)$$

and outside the sphere, we have

$$V_{out}(r, \theta) = -E_0 r \cos \theta + \sum_{l=0}^{\infty} \frac{B_l}{r^{l+1}} P_l(\cos \theta).$$

From the first boundary condition, at $r = R$,

$$V_{in} = V_{out}$$

So

$$\sum_{l=0}^{\infty} A_l R^l P_l(\cos \theta) = -E_0 R \cos \theta + \sum_{l=0}^{\infty} \frac{B_l}{R^{l+1}} P_l(\cos \theta).$$

For $l = 1$, $P_1(\cos \theta) = \cos \theta$. So

$$A_1 R \cos \theta = -E_0 R \cos \theta + \frac{B_1}{R^2} \cos \theta$$

And

$$A_1 R = -E_0 R + \frac{B_1}{R^2}.$$

For $l \neq 1$,

$$A_l R^l = \frac{B_l}{R^{l+1}}.$$

From the second boundary condition

$$\varepsilon \frac{\partial V_{in}}{\partial r} = \varepsilon_0 \frac{\partial V_{out}}{\partial r}$$

we have

$$\varepsilon_r \sum_{l=0}^{\infty} l A_l R^{l-1} P_l(\cos \theta) = -E_0 \cos \theta - \sum_{l=0}^{\infty} \frac{(l+1) B_l}{R^{l+2}} P_l(\cos \theta).$$

For $l \neq 1$,

$$\varepsilon_r l A_l R^{l-1} = -\frac{(l+1) B_l}{R^{l+2}}$$

so we must have
$$A_l = B_l = 0.$$

For $l = 1$
$$\varepsilon_r A_1 = -E_0 - \frac{2B_1}{R^3}.$$

Consider our two equations relating A_1 and B_1,
$$A_1 R = -E_0 R + \frac{B_1}{R^2}$$

and
$$\varepsilon_r A_1 = -E_0 - \frac{2B_1}{R^3}.$$

From the first,
$$A_1 = -E_0 - \frac{B_1}{R^3},$$

and substitution into the second yields
$$\varepsilon_r \left(-E_0 - \frac{B_1}{R^3} \right) = -E_0 - \frac{2B_1}{R^3} \rightarrow B_1 \left(\frac{\varepsilon_r}{R^3} + \frac{2}{R^3} \right) = E_0 (\varepsilon_r - 1)$$

So
$$B_1 = \frac{\varepsilon_r - 1}{\varepsilon_r + 2} R^3 E_0$$

and
$$A_1 = -\frac{3E_0}{\varepsilon_r + 2}.$$

Therefore the potential is
$$V_{\text{in}}(r, \theta) = -\frac{3E_0}{\varepsilon_r + 2} r \cos \theta.$$

Noting that $z = r \cos \theta$,
$$V_{\text{in}} = -\frac{3E_0}{\varepsilon_r + 2} z.$$

The field inside the sphere is uniform and in the same direction as $\vec{E}_0$:
$$\vec{E} = \frac{3}{\varepsilon_r + 2} \vec{E}_0.$$

Bibliography

Byron F W and Fuller R W 1992 *Mathematics of Classical and Quantum Physics* (New York: Dover)
Griffiths D J 1999 *Introduction to Electrodynamics* 3rd edn (Englewood Cliffs, NJ: Prentice Hall)
Griffiths D J 2013 *Introduction to Electrodynamics* 4th edn (New York: Pearson)
Halliday D, Resnick R and Walker J 2010 *Fundamentals of Physics* 9th edn (New York: Wiley)
Halliday D, Resnick R and Walker J 2013 *Fundamentals of Physics* 10th edn (New York: Wiley)
Jackson J D 1998 *Classical Electrodynamics* 3rd edn (New York: Wiley)
Purcell E M and Morin D J 2013 *Electricity and Magnetism* 3rd edn (Cambridge: Cambridge University Press)
Rogawski J 2011 *Calculus: Early Transcendentals* 2nd edn (San Francisco, CA: Freeman)

IOP Concise Physics

Electromagnetism
Problems and solutions
Carolina C Ilie and Zachariah S Schrecengost

Chapter 6

Magnetic fields in matter

Similarly to the electric field in matter and the electric dipoles, when magnetic dipoles are subjected to a magnetic field, they may align and the medium becomes magnetized. Depending on the magnetization, we define different magnetic materials: paramagnets (the magnetization $\vec{M}$ is parallel to the magnetic field $\vec{B}$), diamagnets (the magnetization $\vec{M}$ is opposite to the magnetic field $\vec{B}$), and the special class of materials, ferromagnets, which remain magnetized even after the magnetic field becomes zero.

We would like to mention here that different sources may have different names for $\vec{B}$ and $\vec{H}$. Here $\vec{B}$ is the magnetic field and $\vec{H}$ is simply the H-field. Griffiths refers to $\vec{H}$ as the auxiliary field but we have chosen the H-field to eliminate any confusion. In other books, you may find that $\vec{B}$ is the magnetic flux density, while $\vec{H}$ is the magnetic field.

6.1 Theory

6.1.1 Torque on a magnetic dipole moment

The torque on a magnetic dipole moment $\vec{m}$ in a magnetic field $\vec{B}$ is

$$\vec{N} = \vec{m} \times \vec{B}.$$

6.1.2 Force on a magnetic dipole

The force on a magnetic dipole moment $\vec{m}$ in a magnetic field $\vec{B}$ is

$$\vec{F} = \nabla(\vec{m} \cdot \vec{B}).$$

6.1.3 H-field

Given magnetic field $\vec{B}$ and magnetization $\vec{M}$, the H-field is

$$\vec{H} = \frac{1}{\mu_0}\vec{B} - \vec{M}.$$

with
$$\nabla \times \vec{H} = \vec{J}_{\text{f}},$$
where $\vec{J}_{\text{f}}$ is the free current density. From Stoke's law,
$$\oint_S \vec{H} \cdot d\vec{\ell} = I_{\text{f}_{\text{enc}}},$$
where $I_{\text{f}_{\text{enc}}}$ is the free current.

6.1.4 Linear media

Given H-field $\vec{H}$, the magnetization is given by
$$\vec{M} = \chi_{\text{m}} \vec{H},$$
where χ_{m} is the magnetic susceptibility. The magnetic field is given by
$$\vec{B} = \mu_0 (\vec{H} + \vec{M}) = \mu_0 (1 + \chi_{\text{m}}) \vec{H} = \mu \vec{H},$$
where μ is the magnetic permeability of the material, and μ_0 is the permeability of the vacuum.

6.1.5 Surface bound current due to magnetization $\vec{M}$

Given magnetization $\vec{M}$ and normal vector $\hat{n}$, the surface bound current is
$$\vec{K}_{\text{b}} = \vec{M} \times \hat{n}.$$

6.1.6 Volume bound current due to magnetization $\vec{M}$

Given magnetization $\vec{M}$ and normal vector $\hat{n}$, the volume bound current is
$$\vec{J}_{\text{b}} = \nabla \times \vec{M}.$$

6.2 Problems and solutions

Problem 6.1. Find the force between the two magnetic dipoles below.

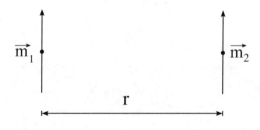

Solution The magnetic field due to dipole $\vec{m}_1$ at $\vec{m}_2$ is given by

$$\vec{B}_{\text{dip},m_1} = \frac{\mu_0 m_1}{4\pi r^3}\left(2\cos\theta\,\hat{r} + \sin\theta\,\hat{\theta}\right),$$

where $\theta = \frac{\pi}{2}$. So

$$\vec{B}_{\text{dip},m_1} = \frac{\mu_0 m_1}{4\pi r^3}\hat{\theta} = -\frac{\mu_0 m_1}{4\pi r^3}\hat{z}.$$

The force on $\vec{m}_2$ due to $\vec{B}_{\text{dip},m_1}$ is

$$\vec{F} = \nabla\left(\vec{m}_2 \cdot \vec{B}_{\text{dip},m_1}\right),$$

where

$$\vec{m}_2 \cdot \vec{B}_{\text{dip},m_1} = m_2\hat{z} \cdot \left(-\frac{\mu_0 m_1}{4\pi r^3}\hat{z}\right) = -\frac{\mu_0 m_1 m_2}{4\pi r^3}.$$

So

$$\vec{F} = \nabla\left(-\frac{\mu_0 m_1 m_2}{4\pi r^3}\right) = -\frac{\mu_0 m_1 m_2}{4\pi}(-3r^{-4})\hat{r} = \frac{3\mu_0 m_1 m_2}{4\pi r^4}\hat{r}.$$

Problem 6.2. Find the force on a dipole located on the axis of an infinitely long cylinder of radius R, rotating at ω and carrying surface charge σ.

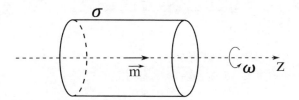

Solution We can think of the rotating cylinder as a solenoid with $nI \to K$. So

$$K = \sigma v = \sigma\omega R$$

and

$$\vec{B} = \mu_0 nI\hat{z} = \mu_0 K\hat{z},$$

which is the magnetic field inside a solenoid. The force is given by

$$\vec{F} = \nabla\left(\vec{m} \cdot \vec{B}\right) = \nabla\left(m\hat{z} \cdot \mu_0 K\hat{z}\right) = \nabla(\mu_0 mK) = 0.$$

This is an example that shows the force on a dipole in a uniform field is zero. Since we can think of a dipole as a current loop, this is equivalent to saying a current loop in a uniform field experiences zero net force.

Problem 6.3. Consider two current loops of radius R whose orientation is depicted below. Find the torque between them and the angle γ that minimizes this torque.

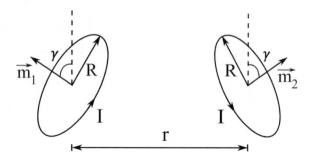

Solution Looking at this from the side, we have

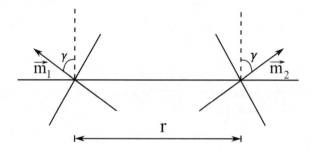

Considering $\vec{m}_1$, the field it produces is

$$\vec{B}_{\text{dip},m_1} = \frac{\mu_0 m_1}{4\pi r^3}(2\cos\theta\,\hat{r} + \sin\theta\,\hat{\theta}),$$

where $\theta = \gamma + \frac{\pi}{2}$, $r = y$, $\hat{r} = \hat{y}$, and $\hat{\theta} = -\hat{z}$. Also, we have

$$m_1 = \pi R^2 I$$

$$\cos\theta = \cos\left(\gamma + \frac{\pi}{2}\right) = -\sin\gamma$$

and

$$\sin\theta = \sin\left(\gamma + \frac{\pi}{2}\right) = \cos\gamma.$$

Using these values,

$$\vec{B}_{\text{dip},m_1} = \frac{\mu_0 \pi R^2 I}{4\pi y^3}(-2\sin\gamma\,\hat{y} - \cos\gamma\,\hat{z}) = -\frac{\mu_0 R^2 I}{4 y^3}(2\sin\gamma\,\hat{y} + \cos\gamma\,\hat{z}).$$

We have $\vec{m}_2$ given by
$$\vec{m}_2 = \pi R^2 I (\sin\gamma\, \hat{y} + \cos\gamma\, \hat{z}).$$
Now the torque is
$$\vec{N} = \vec{m}_2 \times \vec{B}_{\text{dip},m_1} = -\frac{\mu_0 R^2 I}{4y^3}(\pi R^2 I)\Big[(\sin\gamma\,\hat{y} + \cos\gamma\,\hat{z}) \times (2\sin\gamma\,\hat{y} + \cos\gamma\,\hat{z})\Big].$$
Looking at just the cross product term, we have
$$(\sin\gamma\,\hat{y} + \cos\gamma\,\hat{z}) \times (2\sin\gamma\,\hat{y} + \cos\gamma\,\hat{z}) = \begin{vmatrix} \hat{x} & \hat{y} & \hat{z} \\ 0 & \sin\gamma & \cos\gamma \\ 0 & 2\sin\gamma & \cos\gamma \end{vmatrix} = -\sin\gamma\cos\gamma\,\hat{x}.$$
Therefore,
$$\vec{N} = \vec{m}_2 \times \vec{B}_{\text{dip},m_1} = -\frac{\mu_0 \pi R^4 I^2}{4y^3}(-\sin\gamma\cos\gamma\,\hat{x}) = \frac{\mu_0 \pi R^4 I^2 \sin(2\gamma)}{8y^3}\hat{x}.$$
To find the γ that minimizes this, we must differentiate the torque with respect to γ,
$$\frac{\partial N}{\partial \gamma} = \frac{\mu_0 \pi R^4 I^2}{8y^3}\frac{\partial}{\partial \gamma}(\sin(2\gamma)) = \frac{\mu_0 \pi R^4 I^2}{4y^3}\cos(2\gamma).$$
We can find the extreme values by setting this equal to zero. Note we have
$$\frac{\partial N}{\partial \gamma} = 0$$
when
$$\cos(2\gamma) = 0,$$
which is zero when
$$2\gamma = \left(\frac{2n-1}{2}\right)\pi$$
for a positive integer n. Solving for γ we have
$$\gamma = \left(\frac{2n-1}{4}\right)\pi.$$
To find the minimum, we must find the second derivative of the torque. So
$$\frac{\partial^2 N}{\partial \gamma^2} = \frac{\mu_0 \pi R^4 I^2}{4y^3}\frac{\partial}{\partial \gamma}[\cos(2\gamma)] = -\frac{\mu_0 \pi R^4 I^2 \sin(2\gamma)}{2y^3},$$
Substitution of γ yields
$$\beta = -\frac{\mu_0 \pi R^4 I^2}{2y^3}\sin\left[2\left(\frac{2n-1}{4}\right)\pi\right],$$

where a $\beta > 0$ indicates a minimum. Dropping all but the sign and the sine, we have

$$\beta = -\sin\left(n\pi - \frac{\pi}{2}\right) = -\left[\sin(n\pi)\cos\left(-\frac{\pi}{2}\right) + \cos(n\pi)\sin\left(-\frac{\pi}{2}\right)\right] = \cos(n\pi).$$

We have β is positive for $n = 2, 4, 6, \ldots$ so the torque is minimized for

$$\gamma = \left(\frac{2n-1}{4}\right)\pi,$$

when $n = 2, 4, 6, \ldots$ or for any positive integer n,

$$\gamma = \left(\frac{2(2n)-1}{4}\right)\pi = \left(n - \frac{1}{4}\right)\pi = n\pi - \frac{\pi}{4}.$$

Since each dipole moment is at an angle γ, the angle between them is

$$2\gamma = 2n\pi - \frac{\pi}{2}$$

Note the multiple of $2n\pi$ is just the addition of another complete circle. The result that minimizes the torque is a γ that causes the dipoles to be perpendicular to each other.

Problem 6.4. Consider the rotating cylindrical shell in problem 4.6, where the z-axis starts at the left side of the cylinder. Suppose we place a dipole $\vec{m} = m\hat{z}$ at a distance d from the right-hand side of the cylinder, as depicted below. Find the force on the dipole.

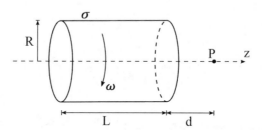

Solution From problem 4.6, the field is given by

$$\vec{B} = \frac{\mu_0 \sigma \omega R}{2}\left[\frac{d+L}{\sqrt{R^2 + (d+L)^2}} - \frac{d}{\sqrt{R^2 + d^2}}\right]\hat{z},$$

where d was the distance from our point to the right-hand side of the cylinder. We can rewrite this considering $d = z - L$. So

$$\vec{B} = \frac{\mu_0 \sigma \omega R}{2} \left[\frac{z}{\sqrt{R^2 + z^2}} - \frac{z - L}{\sqrt{R^2 + (z - L)^2}} \right] \hat{z}.$$

The force is given by

$$\vec{F} = \nabla (\vec{m} \cdot \vec{B})$$

with

$$\vec{m} \cdot \vec{B} = \frac{\mu_0 \sigma \omega R m}{2} \left[\frac{z}{\sqrt{R^2 + z^2}} - \frac{z - L}{\sqrt{R^2 + (z - L)^2}} \right].$$

So

$$\vec{F} = \frac{\mu_0 \sigma \omega R m}{2} \nabla \left[\frac{z}{\sqrt{R^2 + z^2}} - \frac{z - L}{\sqrt{R^2 + (z - L)^2}} \right]$$

and

$$\vec{F} = \frac{\mu_0 \sigma \omega R^3 m}{2} \left[\frac{1}{(R^2 + z^2)^{3/2}} - \frac{1}{[R^2 + (z - L)^2]^{3/2}} \right] \hat{z}.$$

At a distance d from the right-hand side of the cylinder, $z = d + L$. Therefore

$$\vec{F} = \frac{\mu_0 \sigma \omega R^3 m}{2} \left[\frac{1}{[R^2 + (d + L)^2]^{3/2}} - \frac{1}{(R^2 + d^2)^{3/2}} \right] \hat{z}.$$

Problem 6.5. An infinitely long cylinder has a constant magnetization $\vec{M}$ parallel to the axis of the cylinder. Find the magnetic field due to $\vec{M}$ everywhere.

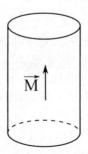

Solution The magnetization is given by

$$\vec{M} = M\hat{z}.$$

The bound volume current $\vec{J}_b$ is

$$\vec{J}_b = \nabla \times \vec{M} = 0$$

since $\vec{M}$ = constant. The bound surface current $\vec{K}_b$ is

$$\vec{K}_b = \vec{M} \times \hat{n} = M\hat{z} \times \hat{s} = M\hat{\phi}.$$

For $s < R$

$$\vec{B} = \mu_0 \vec{M} = \mu_0 M\hat{z}$$

and for $s > R$

$$\vec{B} = 0.$$

Outside, the field is zero ($\vec{B} = 0$ outside a solenoid).

Problem 6.6. A long circular cylinder of radius R has a magnetization $\vec{M} = ks\hat{\phi}$, where k is a constant, s the distance from the axis of the cylinder, and $\hat{\phi}$ the azimuthal unit vector. Find the magnetic field due to $\vec{M}$ for $s < R$ and $s > R$.

Solution Let us first find the bound volume current $\vec{J}_b$ and the bound surface current $\vec{K}_b$. The bound volume current is given by

$$\vec{J}_b = \nabla \times \vec{M} = \frac{1}{s}\frac{\partial}{\partial s}(sM_\phi)\hat{z} = \frac{1}{s}\frac{\partial}{\partial s}(ks^2)\hat{z} = \frac{1}{s}2ks\hat{z} = 2k\hat{z} = \text{const},$$

and the bound surface current is given by

$$\vec{K}_b = \vec{M} \times \hat{n} = ks(\hat{\phi} \times \hat{s})\Big|_{s=R} = -kR\hat{z}.$$

So the bound current flows up the cylinder, and returns down the surface. Let us check that the total current is zero. The total current due to the bound volume current is given by

$$I_{\text{tot},J_b} = \int \vec{J}_b \cdot d\vec{a} = \int J_b \, da = \int_0^R (2k)(2\pi s \, ds) = \frac{4\pi k R^2}{2} = 2\pi k R^2$$

and the total current due to the bound surface current is

$$I_{\text{tot},K_b} = \int K_b \, d\ell = (-kR)2\pi R = -2\pi k R^2.$$

Since they are equal and opposite, the total current is zero. Now we can find the magnetic field using Ampère's law. For $s < R$,

$$\oint_S \vec{B} \cdot d\vec{\ell} = \mu_0 I_{\text{enc}}.$$

The left-hand side is given by
$$\oint_S \vec{B} \cdot d\vec{\ell} = B 2\pi s$$
and enclosed current is given by
$$I_{enc} = \int_0^s J_b \, da = \int_0^s 2k 2\pi s' ds' = 2k\pi s^2.$$
Therefore,
$$\vec{B} = \mu_0 k s \hat{\phi} = \mu_0 \vec{M}.$$
For $s > R$, our enclosed current is zero, so
$$\vec{B} = 0.$$

Problem 6.7. A long cylinder of radius R carries a magnetization $\vec{M} = ks^3 \hat{\phi}$, where k is a constant. Find the magnetic field due to $\vec{M}$ everywhere.

Solution Let us start by finding the bound currents. The volume bound current is given by
$$\vec{J}_b = \nabla \times \vec{M} = \frac{1}{s}\frac{\partial}{\partial s}(s M_\phi)\hat{z} = \frac{1}{s}\frac{\partial}{\partial s}(s k s^3)\hat{z} = \frac{1}{s}\frac{\partial}{\partial s}(k s^4)\hat{z} = \frac{1}{s} 4 k s^3 \hat{z} = 4 k s^2 \hat{z}$$
and the surface bound current is
$$\vec{K}_b = \vec{M} \times \hat{n} = k s^3 (\hat{\phi} \times \hat{s})\Big|_{s=R} = -k R^3 \hat{z}.$$
We can check that the total bound current is zero. From the bound volume current, we have
$$I_{tot,J_b} = \int \vec{J}_b \cdot d\vec{a} = \int_0^R (4k s^2)(2\pi s \, ds) = \frac{8\pi k s^4}{4}\Big|_0^R = 2\pi k R^4$$
and from the bound surface current, we have
$$I_{tot,K_b} = \int K_b \, d\ell = (-k R^3) 2\pi R = -2\pi k R^4.$$
Therefore, the total bound current is zero, $I_b = 0$. Now we can find the field by using Ampère's law. For $s < R$,
$$\oint_S \vec{B} \cdot d\vec{\ell} = \mu_0 I_{enc}.$$
The left-hand side is
$$\oint_S \vec{B} \cdot d\vec{\ell} = B 2\pi s$$

and the enclosed current is

$$I_{\text{enc}} = \int_0^s J_b \, da = \int_0^s 4ks'^2 2\pi s' \, ds' = \int_0^s 8\pi k(s')^3 \, ds' = 2\pi ks^4.$$

Therefore,

$$B 2\pi s = \mu_0 2\pi k s^4 = \mu_0 k s^3$$

And

$$\vec{B} = \mu_0 k s^3 \hat{\phi} = \mu_0 \vec{M}.$$

For $s > R$, we have zero enclosed current. So,

$$\vec{B} = 0.$$

Problem 6.8. A sphere of radius R carries magnetization $\vec{M} = kr\hat{\phi}$. Find the magnetic field inside and outside.

Solution Since there is no free current, $\vec{H} = 0$. Inside, we have magnetization, so

$$\vec{H} = \frac{1}{\mu_0} \vec{B}_{\text{in}} - \vec{M}.$$

So $\vec{B}_{\text{in}}$ is given by

$$\vec{B}_{\text{in}} = \mu_0 \vec{M} = \mu_0 kr\hat{\phi}.$$

Outside, we have no magnetization either, so

$$\vec{B}_{\text{out}} = 0.$$

Problem 6.9. An infinitely long wire carries current I and is surrounded by material, out to radius R, with magnetization $\vec{M} = k\hat{\phi}$. Find the magnetic field for $s < R$ and $s > R$.

Solution For $s < R$, we have

$$\oint_S \vec{H} \cdot d\vec{\ell} = I_{f_{\text{enc}}}$$

with

$$\oint_S \vec{H} \cdot d\vec{\ell} = H 2\pi s$$

and

$$I_{f_{\text{enc}}} = I.$$

Therefore,
$$H = \frac{I}{2\pi s}\hat{\phi}.$$

Using
$$\vec{H} = \frac{1}{\mu_0}\vec{B}_{in} - \vec{M}$$

the magnetic field inside is given by
$$\vec{B}_{in} = \mu_0(\vec{H} + \vec{M}) = \mu_0\left(\frac{I}{2\pi s} + k\right)\hat{\phi}.$$

For $s > R$, we still have $H = \frac{I}{2\pi s}\hat{\phi}$, but we do not have any magnetization. So
$$\vec{B}_{out} = \mu_0\vec{H} = \frac{\mu_0 I}{2\pi s}\hat{\phi},$$

which is exactly what we would expect from a wire carrying current I.

Problem 6.10. An infinitely long wire, of radius R carries magnetization $\vec{M} = ks^2\hat{z}$. At $s = R$, there is a surface current $\vec{K} = K_0\hat{\phi}$. Find the field for $s < R$ and $s > R$.

Solution For $s < R$, both $\vec{M}$ and $\vec{K}$ contribute to the field. We can see the contribution due to $\vec{K}$ is that of a solenoid,
$$\vec{B}_K = \mu_0 K_0 \hat{z}.$$

We also have $\vec{H} = 0$, so
$$\vec{B}_M = \mu_0 ks^2 \hat{z}.$$

Therefore,
$$\vec{B}_{in} = \vec{B}_K + \vec{B}_M = \mu_0(K_0 + ks^2)\hat{z}.$$

For $s > R$, we have zero magnetization, so
$$\vec{B}_M = 0.$$

Also, there is zero field outside of a solenoid, so
$$\vec{B}_K = 0.$$

Therefore,
$$\vec{B}_{out} = 0.$$

Problem 6.11. Find the H-field produced from a current density $\vec{J}_f = J_0 s \hat{z}$ in two ways.

Solution First, we will use

$$\nabla \times \vec{H} = \vec{J}_f.$$

Note, we must have $\vec{H} = H(s)\hat{\phi}$. So

$$\nabla \times \vec{H} = \frac{1}{s}\frac{\partial}{\partial s}[sH(s)]\hat{z} = J_0 s \hat{z}$$

and

$$\frac{\partial}{\partial s}[s\,H(s)] = J_0 s^2.$$

From this, we have

$$H(s) = \frac{1}{3}J_0 s^2 + C.$$

Since there is zero current at $s = 0$, we have $H(0) = 0 \rightarrow C = 0$. Therefore,

$$\vec{H} = \frac{J_0 s^2}{3}\hat{\phi}.$$

Now we will use

$$\oint_S \vec{H} \cdot d\vec{\ell} = I_{f_{enc}},$$

where

$$I_{f_{enc}} = \int J\,da = \int_0^s 2\pi s' J_0 s' ds' = \frac{2\pi J_0 s^3}{3}$$

and

$$\oint_S \vec{H} \cdot d\vec{\ell} = H 2\pi s.$$

Therefore,

$$\vec{H} = \frac{J_0 s^2}{3}\hat{\phi}.$$

As expected from the first method. Note each equation we used is simply Stoke's theorem applied to the other.

Problem 6.12. This problem was inspired by a different problem presented in the Electrodynamics graduate course by Dr Charles Ebner at the Ohio State University in 2002. A sphere of radius R is uniformly polarized with a polarization $\vec{P}$. Within such a sphere, one can show that $\vec{D} = \frac{2}{3}\vec{P}$ and $\vec{E} = -\frac{\vec{P}}{3\varepsilon_0}$. By using the similarity of the equations of electrostatics and magnetostatics, find $\vec{B}$ and $\vec{H}$ within a uniformly magnetized sphere having magnetism $\vec{M}$.

Solution The equivalent equations for electrostatics and magnetostatics are the following

$$\nabla \times \vec{E} = 0$$
$$\nabla \cdot \vec{D} = 0$$
$$\vec{E} = \frac{\vec{D}}{\varepsilon_0} - \frac{\vec{P}}{\varepsilon_0}.$$

For a system with no free current,

$$\nabla \times \vec{H} = 0$$
$$\nabla \cdot \vec{B} = 0$$
$$\vec{H} = \frac{\vec{B}}{\mu_0} - \vec{M}.$$

Comparing the equations, we note that $\vec{E}$ is equivalent to $\vec{H}$, $\vec{E} \Leftrightarrow \vec{H}$; $\frac{\vec{D}}{\varepsilon_0}$ is equivalent to $\frac{\vec{B}}{\mu_0}$, $\frac{\vec{D}}{\varepsilon_0} \Leftrightarrow \frac{\vec{B}}{\mu_0}$; and $\frac{\vec{P}}{\varepsilon_0}$ is equivalent to $\vec{M}$, $\frac{\vec{P}}{\varepsilon_0} \Leftrightarrow \vec{M}$.
Starting with

$$\vec{D} = \frac{2}{3}\vec{P}$$

we can divide both sides by ε_0

$$\frac{\vec{D}}{\varepsilon_0} = \frac{2}{3\varepsilon_0}\vec{P}.$$

Since $\frac{\vec{P}}{\varepsilon_0} \Leftrightarrow \vec{M}$, we have

$$\frac{2}{3\varepsilon_0}\vec{P} \Leftrightarrow \frac{2}{3}\vec{M}.$$

Using $\frac{\vec{D}}{\varepsilon_0} \Leftrightarrow \frac{\vec{B}}{\mu_0}$, we have

$$\frac{\vec{B}}{\mu_0} = \frac{2}{3}\vec{M}.$$

Therefore,

$$\vec{B} = \frac{2\mu_0}{3}\vec{M}.$$

Now we will look at the electric field,

$$\vec{H} \Leftrightarrow \vec{E}.$$

Using $\frac{\vec{P}}{\varepsilon_0} \Leftrightarrow \vec{M}$,

$$\vec{E} = -\frac{\vec{P}}{3\varepsilon_0} = -\frac{\vec{M}}{3}.$$

Therefore,

$$\vec{H} = -\frac{\vec{M}}{3}.$$

Bibliography

Griffiths D J 1999 *Introduction to Electrodynamics* 3rd edn (Englewood Cliffs, NJ: Prentice Hall)
Griffiths D J 2013 *Introduction to Electrodynamics* 4th edn (New York: Pearson)
Halliday D, Resnick R and Walker J 2010 *Fundamentals of Physics* 9th edn (New York: Wiley)
Halliday D, Resnick R and Walker J 2013 *Fundamentals of Physics* 10th edn (New York: Wiley)
Jackson J D 1998 *Classical Electrodynamics* 3rd edn (New York: Wiley)
Purcell E M and Morin D J 2013 *Electricity and Magnetism* 3rd edn (Cambridge: Cambridge University Press)

编辑手记

　　本书是一部英文版的物理学教材,中文书名可译为《电磁学:问题与解法》.

　　本书的作者有两位.一位是:卡罗来纳·C.伊利耶(Carolina C. Ilie),她是纽约州立大学奥斯威戈分校的教授.教授电磁理论近十年,她为学生的考试、小组作业和测验设计了各种问题,并且在过去的两年里,她撰写了有关电磁学和电动力学的两本书.伊利耶博士获得了内布拉斯加大学林肯分校的物理学和天文学博士学位,俄亥俄州立大学的物理学硕士学位以及罗马尼亚的布加勒斯特大学的物理学硕士学位.她在2013年获得了学术和创新活动的普罗沃特奖,并在2016年获得了总统教学卓越奖.她与她的丈夫(也是物理学家)和他们的两个儿子一起生活在纽约中部.

　　另一位是:撒迦利亚·S.施雷森戈斯特(Zachariah S. Schrecengost),一名纽约州立大学的毕业生.他毕业于物理、软件工程和应用数学专业,并获得了学士学位.他上过伊利耶博士的高级电磁理论课程,并且热爱参与这个项目.他为该项目带来了学生学习电动力学的崭新视角,以及一位电动力学和高级数学爱好者的热情和才华.施雷森戈斯特先生曾在锡拉丘兹担任软件工程师,目前正在攻读物理学博士学位.

　　本书的插图画家为茱莉亚·R. D'罗萨路(Julia R. D'Rozario).她于2016年12月毕业于纽约州立大学奥斯威戈分校,在那里获得了物理学学士学位和电影与屏幕研究文学学士学位,并于2016年5月完成了天文学的辅修课程.她上过伊利耶博士的高级电磁理论课程,并在电影界拥有丰富的艺术经验. D'罗萨路女士拥有电子动力学方面的知识以及使用 Inkscape 软件进行绘图的才能. 她未来的目标是读研究生,并继续将她对物理学和电影的热情结合在一起.

正如作者在前言中所介绍的:

> 我们写的这本关于问题和解决方法的书针对的对象是:二年级、三年级、四年级的本科生,他们可能希望研究更多的问题,并在学习时立即获得反馈. 作者强烈推荐读者将大卫·J. 格里菲斯的教科书《电动力学入门》(*Introduction to Electrodynamics*) 作为第一本学习手册,因为它被认为是本科水平中最好的电动力学著作之一. 我们认为这本书可以陪伴那些希望更独立地研究电动力学问题的学生,以便提高他们理解和解决问题的能力,并可以为研究生阶段的学习做好准备. 我们添加了简要的理论注释和公式,对于完整的理论方法,我们建议读者去读格里菲斯的书. 每章的组织方式如下:首先是简要的理论注释,然后是带有解决方案的问题论述. 每章结尾都有简短的文献注记.
>
> 我们计划再写一本关于电动力学的书,这本书将会从讨论麦克斯韦方程和守恒定律开始,然后是电磁波、势、场、辐射和相对论电动力学.
>
> 在这里,我们使用的符号与格里菲斯的相同. 因此,我们将 r 用于从源点 r' 到场点 r 的向量. 请注意 $\hat{r} = \dfrac{r}{r} = \dfrac{r-r'}{|r-r'|}$,就像我们看到的这样,这种表示法大大地简化了复杂的方程式,但是要特别小心符号,特别是如果仅使用草书或使用键盘输入字母的时候. 此外,我们在柱面坐标系中使用相同的符号 s 表示到 z 轴的距离,这与格里菲斯的书中使用的符号相同.
>
> 书中选择的单位是国际单位制 SI 单位. 读者应注意,其他书籍可能使用高斯系统(CGS)或海维赛德—洛伦兹(Heaviside-Lorentz)(HL)系统. 这是每个系统中的库仑力:
>
> 在 SI 系统中
>
> $$\boldsymbol{F} = \frac{1}{4\pi\varepsilon_0} \frac{q_1 q_2}{r^2} \hat{r}$$
>
> 在 CGS 中
>
> $$\boldsymbol{F} = \frac{q_1 q_2}{r^2} \hat{r}$$
>
> 在 HL 系统中
>
> $$\boldsymbol{F} = \frac{1}{4\pi} \frac{q_1 q_2}{r^2} \hat{r}$$

其中一些问题是典型的实践问题，具有提高理解和解决问题能力的教学作用．这里介绍的几个问题是经典的例子，虽然它们出现在各种有关电磁学的本科教科书中，但我们认为，如果在本书中忽略这些问题，那么本书就是不完整的，因为它们是研究电磁学的基础．我们还会提出本质上更普遍的问题，这可能会更具有挑战性．我们试图在两种类型的问题之间保持平衡，希望读者喜欢这种变化，并在创造和解决这些问题时像我们一样兴奋和激动．

本书的版权编辑李丹女士为了使读者能够快速地了解本书的基本内容，特翻译了本书的目录，如下：

前言
致谢
关于作者
1　数学方法
　1.1　理论
　　1.1.1　点积和叉积
　　1.1.2　分离向量
　　1.1.3　变换矩阵
　　1.1.4　梯度
　　1.1.5　散度
　　1.1.6　旋度
　　1.1.7　拉普拉斯算子
　　1.1.8　线积分
　　1.1.9　面积分
　　1.1.10　体积积分
　　1.1.11　梯度的基本定理
　　1.1.12　散度的基本定理(高斯定理，格林定理，散度定理)
　　1.1.13　旋度的基本定理(斯托克斯定理，旋度定理)
　　1.1.14　柱面极坐标
　　1.1.15　球极坐标
　　1.1.16　一维狄拉克 δ 函数
　　1.1.17　矢量场理论
　1.2　问题与解法
　参考文献

2 静电学
 2.1 理论
  2.1.1 库仑定律
  2.1.2 电场
  2.1.3 高斯定律
  2.1.4 E 的旋度
  2.1.5 点电荷分布的能量
  2.1.6 连续分布的能量
  2.1.7 单位体积的能量
 2.2 问题与解法
 参考文献

3 电势
 3.1 理论
  3.1.1 拉普拉斯方程
  3.1.2 解拉普拉斯方程
  3.1.3 一般解
  3.1.4 镜像法
  3.1.5 由偶极子引起的势
  3.1.6 多级膨胀
  3.1.7 单极子矩
 3.2 问题与解法
 参考文献

4 静磁学
 4.1 理论
  4.1.1 磁力
  4.1.2 载流导线上的力
  4.1.3 体电流密度
  4.1.4 连续方程
  4.1.5 毕奥—萨伐尔定律
  4.1.6 B 的散度
  4.1.7 安培定律
  4.1.8 矢势
  4.1.9 磁偶极矩
  4.1.10 偶极矩引起的磁场
 4.2 问题与解法

参考文献
5　物质中的电场
　5.1　理论
　　5.1.1　电场中原子的感生偶极矩
　　5.1.2　电场对偶极子的力矩
　　5.1.3　偶极子上的力
　　5.1.4　电场中偶极子的能量
　　5.1.5　极化 $\boldsymbol{P}$ 引起的表面束缚电荷
　　5.1.6　极化 $\boldsymbol{P}$ 引起的体积束缚电荷
　　5.1.7　极化 $\boldsymbol{P}$ 引起的势
　　5.1.8　电位移
　　5.1.9　电位移的高斯定律
　　5.1.10　线性电介质
　　5.1.11　电介质系统中的能量
　5.2　问题与解法
　参考文献
6　物质中的磁场
　6.1　理论
　　6.1.1　磁偶极矩的力矩
　　6.1.2　磁偶极子的力
　　6.1.3　H 场
　　6.1.4　线性介质
　　6.1.5　由磁化 $\boldsymbol{M}$ 引起的表面束缚电流
　　6.1.6　由磁化 $\boldsymbol{M}$ 引起的体积束缚电流
　6.2　问题与解答
　参考文献

　　本书的第一章介绍了一些数学方法,如:狄拉克 δ 函数等.哈尔滨师范大学物理与电子工程学院,先进功能材料与激发态黑龙江省重点实验室的徐玲玲,哈尔滨工业大学物理系的赵永芳和井孝功三位教授2010年曾从狄拉克 δ 函数 $\delta(x)$ 的定义出发,说明其在 $x=0$ 处无定义,$\delta(x)$ 具有 x 倒数的量纲,并且具有 x 算符本征函数的物理含意,澄清了量子力学教材中的一些模糊认识.

众所周知,为了解决连续谱的本征波函数不能归一化的问题,狄拉克[①]引入了一个实函数 $\delta(x)$,称之为狄拉克 δ 函数(以下简称为 δ 函数),从而使得连续谱的本征波函数可以规格化为 δ 函数.

1 δ 函数的定义

狄拉克 δ 函数的原始定义为:以坐标 x 为自变量,并且同时满足下列两个条件的实函数 $\delta(x)$ 称之为 δ 函数

$$\begin{cases} \delta(x)=0, & x\neq 0 \\ \int_{-\infty}^{+\infty}\delta(x)\mathrm{d}x=1 \end{cases} \tag{1}$$

除了上述原始的定义之外,δ 函数还有另外一些表述方式.
(1)导数形式.
若已知阶梯函数为

$$\theta(x)=\begin{cases} 0, & x<0 \\ 1, & x>0 \end{cases} \tag{2}$$

则容易验证 $\theta(x)$ 的一阶导数 $\theta'(x)$ 刚好满足式(1)的两个条件,于是有

$$\delta(x)=\theta'(x) \tag{3}$$

(2)积分形式.
将 $\delta(x)$ 向波数 k 的本征波函数展开,即

$$\delta(x)=\frac{1}{\sqrt{2\pi}}\int_{-\infty}^{+\infty}\varphi(k)\exp(\mathrm{i}kx)\mathrm{d}k \tag{4}$$

展开系数为

$$\varphi(k)=\frac{1}{\sqrt{2\pi}}\int_{-\infty}^{+\infty}\delta(x)\exp(-\mathrm{i}kx)\mathrm{d}x=\frac{1}{\sqrt{2\pi}} \tag{5}$$

将式(5)代入式(4),立即得到 $\delta(x)$ 的一种最常用的形式

$$\delta(x)=\frac{1}{2\pi}\int_{-\infty}^{+\infty}\exp(\mathrm{i}kx)\mathrm{d}k \tag{6}$$

(3)极限形式.
设 k_0 为与 k 同量纲的实常数,对式(6)右端积分,还可以得到 $\delta(x)$ 的另一种表达式

$$\delta(x)=\frac{1}{2\pi}\lim_{k_0\to\infty}\int_{-k_0}^{k_0}[\cos(kx)+\mathrm{i}\sin(kx)]\mathrm{d}x=\lim_{k_0\to\infty}\frac{\sin(k_0 x)}{\pi x} \tag{7}$$

---

① 狄拉克.量子力学原理[M].陈咸亨,译.北京:科学出版社,1965.

容易验证,上述 3 种表述形式皆满足 δ 函数的原始定义中式(1)的两个要求.应该特别强调,所有的表达式都不能确切给出 $\delta(x)$ 在 $x=0$ 处的函数值,或者说,在 $x=0$ 处 $\delta(x)$ 没有定义.在这个意义上讲,$\delta(x)$ 并不是一个通常意义下的函数.如果一定要追究 $\delta(0)$ 的取值,也只能说当 $x\to 0$ 时,$\delta(x)\to\infty$.正是这个原因使得数学家们在很长的时间内不愿意接受它.

在有些量子力学的教材中,将 $\delta(x)$ 简单地写成

$$\delta(x)=\begin{cases}0, & x\neq 0 \\ \infty, & x=0\end{cases} \tag{8}$$

上述的写法曲解了 δ 函数的原始定义.由 δ 函数满足的一个基本关系式

$$x\delta(x)=0 \tag{9}$$

可以看出:如果按照式(8)的表述,当 $x=0$ 时,式(9)左端会出现不定式 $0\times\infty$[①],而不是 0,显然,式(8)的表述不妥.

2 δ 函数的图形

由于 $\delta(x)$ 具有特殊的性质,所以不能直接画出它的图形.在一些量子力学教材中,将 $\delta(x)$ 的图形绘制为一条左右对称且非常靠近纵轴的曲线.这种画法并不准确,很容易造成误解.

汤川秀树[②]曾指出,δ 函数是矩形函数的极限情况,具体的做法是引入一个含有参变量 ε 的矩形函数

$$D(x,\varepsilon)=\begin{cases}0, & |x|\geqslant\dfrac{\varepsilon}{2} \\ \varepsilon^{-1}, & |x|<\dfrac{\varepsilon}{2}\end{cases} \tag{10}$$

其中 ε 是一个小的正数.上式表明 $D(x,\varepsilon)$ 是一个长度为 ε、高度为 ε^{-1} 的矩形函数,该函数对纵轴是左右对称的,并且矩形的面积为无量纲的常数 1.显然,$D(x,\varepsilon)$ 是一个处处有定义的通常意义下的函数.

矩形函数 $D(x,\varepsilon)$ 如图 1 所示.

由 δ 函数的定义可知,当 $\varepsilon\to 0$ 时,上述矩形函数变成 δ 函数,即

$$\lim_{\varepsilon\to 0}D(x,\varepsilon)=\delta(x) \tag{11}$$

当 $\varepsilon\to 0$ 时,图 1 的极限情况就是 δ 函数的图形.

---

① 《数学手册》编写组.数学手册[M].北京:高等教育出版社,1990.
② 汤川秀树.量子力学(1)[M].阎寒梅,张帮固,译.北京:科学出版社,1991.

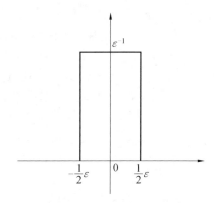

图 1　函数 $D(x,\varepsilon)$ 的矩形函数图形表示

3　δ 函数的量纲

对于以坐标 x 为自变量的 δ 函数 $\delta(x)$ 而言,由式(1)中的第 2 个式子可知,$\delta(x)$ 必须具有长度倒数的量纲 $[L^{-1}]$,不然,等式右端不可能为无量纲的常数 1. 进而可知,三维坐标的 δ 函数 $\delta^3(r)$ 的量纲为 $[L^{-3}]$. 推广到以任意力学量为自变量的情况,例如,对于以动量 p 为自变量的 δ 函数 $\delta^3(p)$ 而言,其量纲为 $[M^{-3}L^{-3}T^3]$.

下面用两个例子来说明 δ 函数量纲的重要性.

(1)单色平面波.

在一维坐标表象下,波数为 $k(-\infty<k<\infty)$ 的状态为

$$\psi_k(x) = \frac{1}{\sqrt{2\pi}} \exp(ikx) \tag{12}$$

在此状态下,坐标的取值概率密度为

$$W_k(x) = |\psi_k(x)|^2 = \frac{1}{2\pi} \tag{13}$$

由坐标取值概率密度的定义可知,$W_k(x)$ 应该具有量纲 $[L^{-1}]$,而式(13)中的 $W_k(x)$ 却是一个无量纲的常数. 出现这种情况的原因是 $\psi_k(x)$ 在这里不能归一化,只能利用式(6)对其进行规格化,即

$$\int_{-\infty}^{+\infty} \psi_{k'}^*(x) \psi_k(x) dx = \frac{1}{2\pi} \int_{-\infty}^{+\infty} \exp[i(k-k')x] dx = \delta(k-k') \tag{14}$$

由于 $\delta(k-k')$ 具有长度的量纲 $[L]$,所以 $W_k(x) = |\psi_k(x)|^2$ 是无量纲的常数,从而具有概率的物理意义,而不是通常意义下的概率密度.

(2)δ 函数位势.

在一些量子力学教材中,将 δ 函数位势写成

$$V(x) = V_0 \delta(x) \tag{15}$$

其中 V_0 是实常量,当 $V_0 > 0$ 时为 δ 势垒,当 $V_0 < 0$ 时为 δ 势阱。如果不再对 V_0 的量纲做特殊的说明,很容易将其误解为具有能量量纲 $[L^2 M T^{-2}]$。实际上,由于 $\delta(x)$ 具有量纲 $[L^{-1}]$,为保证 $V(x)$ 具有能量量纲,V_0 必须具有量纲 $[L^3 M T^{-2}]$。

对于质量为 m,处于 δ 势阱中的粒子而言,由式(15)求得的参量本征值为

$$E = -\frac{mV_0^2}{2\hbar^2} \tag{16}$$

显然,只有当 V_0 具有力的量纲时,式(16)右端才具有能量量纲。

4 δ 函数的含义

在自身表象下,坐标算符 x 满足的本征方程为

$$x\psi_{x_0}(x) = x_0 \psi_{x_0}(x) \tag{17}$$

其中,本征值 x_0 可以在正负无穷之间连续取值,$\psi_{x_0}(x)$ 为其相应的本征波函数。式(17)可以改写成

$$(x - x_0)\psi_{x_0}(x) = 0 \tag{18}$$

由 δ 函数的性质式(9)可知

$$\psi_{x_0}(x) = \delta(x - x_0) \tag{19}$$

于是,$\delta(x - x_0)$ 就具有了坐标算符 x 对应本征值 x_0 的本征波函数的物理含义。

另外,如果将 $\delta(x - x_0)$ 视为算符,在任意状态 $\psi(x)$ 下其平均值为

$$\overline{\delta(x - x_0)} = \int_{-\infty}^{+\infty} \psi^*(x) \delta(x - x_0) \psi(x) \mathrm{d}x = |\psi(x_0)|^2 \tag{20}$$

式(20)表明,$\delta(x - x_0)$ 在任意状态 $\psi(x)$ 下的平均值皆等于坐标在 x_0 处的取值概率密度,或者说,在平均值的意义下,$\delta(x - x_0)$ 具有坐标概率密度算符的物理含义。进而还可以利用 $\delta(x - x_0)$ 定义概率流密度,就不在这里讨论了。

综上所述,原本人为引入的一个 δ 函数,如今已经具有了双重的物理意义:作为波函数,它是坐标算符的本征函数,它的模方表示体系坐标的取值概率密度;作为算符,它是体系坐标的取值概率密度算符。这种集波函数与算符的功能于一身的奇妙现象的出现,必将加深对波函数与算符这两个基本概念的理解。

本书是一部大学物理教材，但其中涉及的数学工具深入探索下去便可抵达近代数学的领域，比如施瓦兹的广义函数论．

施瓦兹（Schwartz, Laurent, 1915—2002），法国人．1915年3月5日生于巴黎．在中学学习时，先是热衷于学习拉丁文和希腊文，后来兴趣转向了数学．1934年考入了法国高等师范学校，学习了当时的现代数学，如勒贝格积分、单复变函数、偏微分方程、现代概率论等．1937年毕业，并取得了教师资格．施瓦兹在大学期间遇到了现代概率论的主要奠基人莱维（1886—1971，后成为他的岳父），这对他的学术道路产生过重大影响．莱维指导过他写概率论方面的论文．1937年至1940年当过三年兵．1943年获博士学位．1944年在格林兹布当讲师．1945年到了布尔巴基学派活动中心南锡，在南锡大学任教授．

施瓦兹在泛函分析、偏微分方程、概率论等领域均做出了重要贡献．但是，最主要的贡献是创立了广义函数（分布）论．人们对函数概念的认识是不断深化的．19世纪随着对数学物理方程的深入研究，出现了许多奇异函数，如无处可微的连续函数；到20世纪，物理学家又提出了一些新函数，如狄拉克提出的"怪"函数——δ函数．施瓦兹早在大学读书期间就考虑过如何将函数概念加以推广，使之能"容下"诸如δ函数等函数的问题．他首先认真研究了费歇尔空间的对偶理论；其次系统地总结了当时许多数学家关于广义函数的零星的想法和理论；在这个基础上，于1945年和1946年间先后发表了4篇关于广义函数的论文，建立了广义函数论的完整体系．他在建立广义函数论理论体系过程中所起的作用，与牛顿、莱布尼兹在建立微积分过程中所起的作用有些类似．现代的几乎大部分数学新分支都建立在这个基础上．因此，他在这方面的贡献在数学史上的意义是深远的．施瓦兹于1950年和1951年发表了两卷本专著《广义函数论》．这是广义函数理论的经典著作．

1950年施瓦兹荣获第二届菲尔兹奖．

再比如本书的第四章介绍的毕奥—萨伐尔定律．

毕奥—萨伐尔定律是电磁学中一个重要的定律，它是法国科学家毕奥和萨伐尔合作研究发现载流长直导线对磁极的作用反比于距离的实验结果，并确定了电流对磁极的作用力为横向力，该定律是由实验方法得到的，为了能够更好地掌握毕奥—萨伐尔定律的建立过程，集宁师范学院的呼和满都拉，冀文慧，杨洪涛和胡晓颖四位教授2014年根据电磁场理论，应用电磁场变换的关系式推导运动电荷产生的磁场，进而得出了经典物理学上的重要电磁规律——毕奥—萨伐尔定律．在该条件下求出的毕奥—萨伐尔定律不仅适用条件清晰明了，同时为应用相对性原理解决问题做了充足的准备．

19世纪以前，人们一直认为电学和磁学的研究一直都是独立地发展着的，尽管"顿牟掇芥、磁石引针"的描述，以及存在平方反比关系的电场力和磁场力都在不同程度上反映了电磁场的相似性,但是相似和相等并不成等价关系.直到1820年,丹麦的物理学家奥斯特在前人研究发现的基础上,通过大量的演示实验和研究分析,总结出了电流和磁针间存在力的作用,并做了相关报告,总结了他60次的电流磁效应的实验[①],电与磁的发展才进入了新的时期.正在瑞士访问的法国科学院院士阿拉果听到这一重要消息,立即带着这一新闻回到法国,并在法国科学院报告和演示了奥斯特的重大发现.由于此前一直深受库仑的影响,报告一出来,就在法国的科学界引起了强烈的轰动.法国科学家立即重新审视电和磁的关系,并着手进行试验.很快,安培就在科学院会议上对相关的3篇论文做了报告.报告会上做了演示实验,证明通电螺线管能像磁铁一样相互吸引[②],同时毕奥和萨伐尔合作对电的磁效应展开着定量研究,1820年9月30日,两人将第一个实验结果发表,载流长直导线到磁极距离与其作用力成反比的结果.这是人类第一次得到电流磁效应的定量结果.在这种背景下建立起来的不仅是毕奥-萨伐尔定律,同时肯定了电和磁的联系.

毕奥-萨伐尔定律虽然十分重要,但是日常教学中对该定律的讲解,通常是把注意力主要集中在应用该定律解题和计算上,这种引入方式不仅忽略了建立定律的物理内涵、研究方法和创新精神,而且没有深刻反映定律的建立过程.在这里,我们先简单介绍定律的建立背景,了解它的实验原理和方法.再应用相对性原理和电磁场变换公式,具体地阐明如何从理论上推导出该定律.最后用根据安培定律得出的毕奥-萨伐尔定律对上一种方法进行验证.

毕奥-萨伐尔定律的建立是不可争论的事实,它在电磁学中的地位等同于库仑定律在静电学中的重要地位.现在的技术发明,凡是与电相联系的都会有磁的参与,如手机、电视、电脑、电动机等,各种电子设备都应用到电磁的转换.该定律的建立对人类关于电磁现象的认识做出了突出的贡献,并对电磁学的发展具有里程碑式的意义.

1 建立毕奥-萨伐尔定律的实验过程

由于受库仑定律的深刻影响,科学家一致认为电和磁之间没有关系,并对磁和电分别进行着研究.当奥斯特发现电流磁效应的消息传到法国科学界后,相隔一周,安培就取得了重要的研究成果.与此同时,毕奥和萨

---

① 周武雷.毕奥-萨伐尔定律的教育价值评述.中国科教创新导刊,2009:81.
② 宋德生,李国栋.电磁学发展史.南宁:广西人民教育出版社,1996:135-136.

伐尔进行了相关的电磁的实验,并且在法国科学院会议上做了重要报告.他们观察到:磁针的南极和北极被载流长直导线施加的力都反比于磁极与导线之间的距离,这个实验结果是人类首次对电流磁效应的定量研究.

1.1 毕奥－萨伐尔定律的实验原理

毕奥和萨伐尔完美地应用实验方法建立了毕奥－萨伐尔定律,其整个实验原理可以分为3个方面:首先,在测量小磁针的受力状况时应用了磁针振荡周期法,求出磁力反比于振荡同期的平方,巧妙地通过振荡周期间接地测量载流直导线产生的磁场作用在磁极上的力;其次,为了消除地磁场的影响,他们采用补偿法避免了物理学实验中地磁场对磁针磁化产生的实验误差,提高了实验的精确度;最后,根据具有对称性的弯折导线作用在磁极上的力的实验,求解作用力的大小与弯折角度的关系,并且导出定量的公式.那么,从一些特殊的电流产生的磁场导出电流源产生磁场的一般公式,就是毕奥－萨伐尔定律实验过程的实质[①].

1.2 毕奥－萨伐尔定律的实验步骤

毕奥和萨伐尔的实验设计思想十分独特,具有创造性,其基本过程分为两个实验来完成.

(1)实验1.

步骤一:测量小磁针受力与振荡周期的关系.取一无限长载流直导线和一枚小磁针,将小磁针悬挂于导线正上方,二者之间有一定的距离.磁针的两端相当于两个极,根据小磁针南北极的受力状况,毕奥和萨伐尔得出这样的结论,一条无限长载流直导线作用在南北磁分子上的作用力都垂直于该分子到导线的距离.从这个结论出发,两人用了周期振荡法和转动定律.解得小磁针振荡周期与力的关系为

$$F \propto \frac{1}{T^2}$$

步骤二:测量小磁针受力与磁针两端到导线距离的关系[②].为了小磁针开始试验时水平地静止,需要消除地球磁场对实验的影响再进行实验.选取磁针距离导线30 mm处作为标准距离,然后每改变一次间距测量的周期,记录测量的力与在标准距离测量的力的比值.重复多次测量、记录,观察得到磁针振荡周期的平方与距离成正比.结合以上结果可知:由于小

---

① 穆良柱,陈熙谋.毕奥－萨伐尔定律建立过程中的数学分析.大学物理,2008(11):27.
② 王较过.毕奥－萨伐尔定律的建立过程.四川师范大学学报(自然科学版),2001(6):614-617.

磁针受到无限长载流直导线所产生磁场的总作用力与磁极到导线的距离是反比关系.

(2)实验 2.

步骤一：计算磁极受到整个长直导线所给的力.一电流微元是在长直导线任意一点选取的,它会给磁极很小的作用力为

$$\mathrm{d}F = \frac{ki\mathrm{d}\theta}{r\sin\theta}f(\theta)$$

从该式可以看出力是关于 θ 角变化的函数,根据角的变化范围 $0\sim\pi$ 积分得到

$$F = \frac{2ki}{d}\int_0^{\frac{\pi}{2}}f(\theta)\mathrm{d}\theta$$

步骤二：将长直导线弯折,分别记录弯折导线产生的作用力与长直导线产生的作用力的比值,进而确定函数 $f(\theta)$ 的具体形式为 $\sin\theta$.将 $\sin\theta$ 代入

$$F = \frac{2ki}{d}\int_0^{\frac{\pi}{2}}f(\theta)\mathrm{d}\theta$$

求得毕奥－萨伐尔定律的微分形式

$$\mathrm{d}F = \frac{ki\mathrm{d}s}{r^2}\sin\theta$$

现在的常用形式是

$$\mathrm{d}\boldsymbol{B} = \frac{kI\mathrm{d}\boldsymbol{l}}{r^2}\sin\theta$$

毕奥－萨伐尔定律的建立,不仅适用于计算任意形状的稳恒电流产生的磁感应强度,并且反映了电与磁的定量关系.根据毕奥和萨伐尔的实验,法国的数学家、物理学家拉普拉斯遵循将粒子间的引力、斥力与一切物理现象的转化关系,从数学上推导出每个电流元施加在磁极上的作用力的规律,所以有些书中也把该定律称为毕奥－萨伐尔－拉普拉斯定律[①].

2　毕奥－萨伐尔定律的理论推导

在学习毕奥－萨伐尔定律的过程中,掌握该定律的理论推导是十分必要的,它为定量求解磁学问题提供依据.例如在求解载流导线产生的磁场、载流圆线圈产生的磁场、载流螺线管产生的磁场等许多问题中都有广

---

① 王较过.毕奥－萨伐尔定律的建立过程.四川师范大学学报(自然科学版),2001(6):614-617.

泛的应用.通过电磁学的相关知识,发现毕奥—萨伐尔定律可以从多方面推理得到.下面根据相对论条件下的电磁场变换公式进行推导毕奥—萨伐尔定律.这样不仅丰富了电磁统一理论,而且毕奥—萨伐尔定律的正确性也得到了验证.

2.1 电磁场的变换关系

从洛伦兹力协变性和电荷的不变性出发所导出的在不同惯性系之间的电磁场变换的一般公式

$$\begin{cases} \boldsymbol{E}_{/\!/} = \boldsymbol{E}'_{/\!/} \\ \boldsymbol{B}_{/\!/} = \boldsymbol{B}'_{/\!/} \\ \boldsymbol{E}_\perp = \gamma(\boldsymbol{E}' - \boldsymbol{v} \times \boldsymbol{B}')_\perp \\ \boldsymbol{B}_\perp = \gamma\left(\boldsymbol{B}' + \dfrac{\boldsymbol{v}}{c^2} \times \boldsymbol{E}'\right)_\perp \end{cases} \tag{1}$$

如上形式可以写为

$$\begin{cases} E_x = E'_x \quad B_x = B'_x \\ E_y = \gamma(E'_y - \boldsymbol{v} \times B'_y) \\ B_y = \gamma\left(B'_y + \dfrac{\boldsymbol{v}}{c^2} \times E'_y\right) \\ E_z = \gamma(E'_z - \boldsymbol{v} \times B'_z) \\ B_z = \gamma\left(B'_z + \dfrac{\boldsymbol{v}}{c^2} \times E'_z\right) \end{cases} \tag{2}$$

如图 1 所示,结合运动的电荷产生的磁场得到,根据右手定则得到

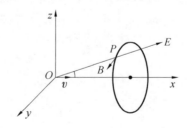

图 1　运动电荷的电磁场

$$\begin{cases} E_x = E'_x \qquad B_x = B'_x \\ E_y = \gamma(E'_y + v \times B'_y) \\ B_y = \gamma(B'_y - \dfrac{1}{c^2} v E'_z) \\ E_z = \gamma(E'_z - v B'_y) \\ B_z = \gamma(B'_z + \dfrac{1}{c^2} v E'_y) \end{cases} \qquad (3)$$

式子中的 $\gamma = \dfrac{1}{\sqrt{1-\beta^2}}$，$\beta = \dfrac{v}{c}$，其中 v 是系 S' 相对于系 S 的运动速度.

2.2 毕奥－萨伐尔定律的推导过程

通过对电磁学知识的学习，了解到电荷或变化的磁场能够激发电场，其中静电场是由静止的点电荷激发而来的. 但是磁场不会由静止电荷激发，只有电流或运动的电荷才能够产生磁场. 也就是说磁场是由运动电荷激发出来的，也是由变化的电场激发出来的. 从这一结论就可以看到电和磁的紧密联系. 下面我们就可以利用电学知识解决磁学问题，得出相关的磁学规律.

2.2.1 变化的电场激发磁场

一般人们认为磁场是被位移电流激发的. 但是这种单一的说法是不完整的，磁场是有"源"场的. 而实质上只有运动电荷才能够激发磁场（运动电荷就是磁场的"源"）. 由于知道磁场的"源"，那么对于惯性系运动和静止是相对的. 设想一下，电荷在某一状态的参考系下不存在，是否在其他状态的参考系下也不存在呢？下面就带着这一问题来推导磁场公式. 首先在推导之前要了解变化电场的具体形式，及其与磁场的关系.

假设存在两个惯性系 S 和 S'，一带电荷量为 q 的点电荷，在系 S 中以 $v = v i$ 的速度做匀速运动. 其中惯性系 S' 也以相同的速度相对于惯性系 S 运动. 此时刻电荷 q 正处在惯性系 S' 的原点. 如图 2 所示，讨论不同惯性系下的电磁场.

坐标系 S' 中，电荷 q 此时对于该系是静止状态，那么电荷在系 S' 中只能产生静电场，而不产生磁场. 在空间中任意一点 $P(x', y', z')$ 处，其电磁场强度分别为

$$\boldsymbol{E}' = \dfrac{q}{4\pi\varepsilon_0 r^3} \boldsymbol{r} \qquad (4)$$

$$\boldsymbol{B}' = \boldsymbol{0} \qquad (5)$$

将电场强度在该系中的 3 个分量根据下图（图 3）表示出来. 将其矢量

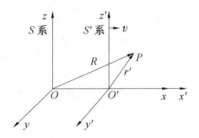

图2 电荷在两个相对运动的惯性系下运动

式分解成分量式

$$\begin{cases} E'_x = E'\cos\beta\cos\alpha = \dfrac{q}{4\pi\varepsilon_0 r'^3} r'\cos\beta\cos\alpha \\ E'_y = E'\sin\beta = \dfrac{q}{4\pi\varepsilon_0 r'^3} r'\sin\beta \\ E'_z = E'\cos\beta\sin\alpha = \dfrac{q}{4\pi\varepsilon_0 r'^3} r'\cos\beta\sin\alpha \end{cases} \quad (6)$$

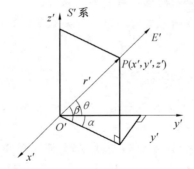

图3 相对参考系静止的电荷的电场

如图3所示,将三角函数式写为坐标式

$$\begin{cases} E'_x = \dfrac{q}{4\pi\varepsilon_0 r'^3} x' \\ E'_y = \dfrac{q}{4\pi\varepsilon_0 r'^3} y' \\ E'_z = \dfrac{q}{4\pi\varepsilon_0 r'^3} z' \end{cases} \quad (7)$$

根据式(3)和式(5)得到

$$E_x = E'_x, \quad E_y = \gamma E'_y, \quad E_z = \gamma E'_z \quad (8)$$

由洛伦兹变换

$$\begin{cases} x' = \gamma(x - vt) \\ y' = y \\ z' = z \end{cases} \quad (9)$$

将式(7)和式(9)代入式(8),把在 S 系中场的分量的时空坐标表示出来

$$\begin{cases} E_x = \dfrac{1}{4\pi\varepsilon_0} \cdot \dfrac{q\gamma(x-vt)}{[\gamma^2(x-vt)^2 + y^2 + z^2]^{\frac{3}{2}}} \\ E_y = \dfrac{1}{4\pi\varepsilon_0} \cdot \dfrac{q\gamma y}{[\gamma^2(x-vt)^2 + y^2 + z^2]^{\frac{3}{2}}} \\ E_z = \dfrac{1}{4\pi\varepsilon_0} \cdot \dfrac{q\gamma z}{[\gamma^2(x-vt)^2 + y^2 + z^2]^{\frac{3}{2}}} \end{cases} \quad (10)$$

观察在系 S 时空坐标中,空间中电荷在运动,时空在变化,电场也在发生改变.当时间处在零时刻 $t=0$,电荷又恰好位于原点,则场强变为

$$\begin{cases} E_x = \dfrac{1}{4\pi\varepsilon_0} \cdot \dfrac{q\gamma x}{[\gamma^2 x^2 + y^2 + z^2]^{\frac{3}{2}}} \\ E_y = \dfrac{1}{4\pi\varepsilon_0} \cdot \dfrac{q\gamma y}{[\gamma^2 x^2 + y^2 + z^2]^{\frac{3}{2}}} \\ E_z = \dfrac{1}{4\pi\varepsilon_0} \cdot \dfrac{q\gamma z}{[\gamma^2 x^2 + y^2 + z^2]^{\frac{3}{2}}} \end{cases} \quad (11)$$

合矢量 $\boldsymbol{E}$ 的大小

$$|\boldsymbol{E}| = \sqrt{E_x^2 + E_y^2 + E_z^2} = \sqrt{\dfrac{q^2}{(4\pi\varepsilon_0)^2} \dfrac{\gamma^2(x^2+y^2+z^2)}{(\gamma^2 x^2 + y^2 + z^2)^3}}$$

将 $\gamma = \dfrac{1}{\sqrt{1-\beta^2}}$ 代入上式,所以

$$|\boldsymbol{E}| = \dfrac{q}{4\pi\varepsilon_0 r^2} \cdot \dfrac{1-\beta^2}{(1-\beta^2\sin^2\theta)^{\frac{3}{2}}} \quad (12)$$

式(11)和式(12)是在相对论的参考系空间中电场强度的分量形式和合矢量形式.

匀速运动情形下的电荷,既能产生变化的电场,又能激发磁场.如何求得运动电荷产生的磁场,这就需要应用电磁场变换公式(3),(5)和式(8),得到以速度 $\boldsymbol{v}=v\boldsymbol{i}$ 运动的电荷在空间磁感应强度是

$$\begin{cases} B_x = B'_x = 0 \\ B_y = \gamma\left(0 - \dfrac{v}{c^2}E'_z\right) = -\gamma\dfrac{v}{c^2}E'_z = -\dfrac{v}{c^2}E_z \\ B_z = \gamma\left(0 + \dfrac{v}{c^2}E'_y\right) = \gamma\dfrac{v}{c^2}E'_y = \dfrac{v}{c^2}E_y \end{cases} \quad (13)$$

可以写作合矢量形式

$$B=\frac{1}{c^2}v\times E \qquad (14)$$

由此,在 S 系中观察到的运动电荷激发的磁场是电场的一种相对论效应. 式(13),(14)就表示磁场的大小.

2.2.2 毕奥-萨伐尔定律的导出

将 $t=0$ 时刻的式(12)代入式(14),磁场强度变化为

$$B=\frac{1}{4\pi\varepsilon_0 c^2 r^2}\cdot\frac{qv(1-\beta^2)\sin\theta}{(1-\beta^2\sin^2\theta)^{\frac{3}{2}}} \qquad (15)$$

前面提到 $\beta=\dfrac{v}{c}$,速度 v 相对于光速非常小,那么 $\beta\ll 1$,即 $\beta\rightarrow 0$ 能够将式(15)化简成

$$B=\frac{1}{4\pi c^2\varepsilon_0}\cdot\frac{qv\sin\theta}{r^2} \qquad (16)$$

光学中的真空介电常数 ε_0 和真空磁导率 μ_0 存在如下关系[①]

$$c=\frac{1}{\sqrt{\varepsilon_0\mu_0}}$$

所以

$$\varepsilon_0 c^2=\frac{1}{\mu_0} \qquad (17)$$

将式(17)代入式(16)有

$$B=\frac{\mu_0}{4\pi}\cdot\frac{qv\sin\theta}{r^2}$$

可以写作矢量形式

$$B=\frac{\mu_0}{4\pi}\frac{q}{r^3}v\times r \qquad (18)$$

从上式可知磁场的方向垂直于 v 与 r 所确定的平面,但这只适用于宏观低速($v\ll c$)的状态下,这就是匀速运动的点电荷产生的磁场近似公式. 根据金属导电的经典电子理论,金属导体中的电流是自由电子定向运动形成的,所以恒定电流所产生的磁场可以看作所有匀速运动电荷产生的磁场的总和. 由此,可以将运动电荷产生的磁场过渡到电流产生的磁场.

假设 n 个电子被包含于单位体积内的导体中,每个电子的电荷量是 $-e$,且以匀速定向 v 运动,在长度为 l 的闭合环路导体上选取 dl 导体元.

---

① 姚启钧.光学教程.北京:高等教育出版社,2008:9.

则导体元带电荷量为 $dq = ensdl$，代入电荷量为 dq 产生的磁场强度公式 $d\boldsymbol{B} = \dfrac{dq}{r^3}\boldsymbol{v} \times \boldsymbol{r}$ 中有

$$d\boldsymbol{B} = \dfrac{ensdl}{r^3}\boldsymbol{v} \times \boldsymbol{r} \tag{19}$$

导体中电流强度 I 的大小与微观量 n, e, v 的关系是 $I = n(-e)sv$。又由于电子带负电，电流的方向与电子运动的方向相反，所以 $d l \boldsymbol{v} = -v d \boldsymbol{l}$，那么式(19)就可以改写为[1]

$$d\boldsymbol{B} = \dfrac{\mu_0}{4\pi} \cdot \dfrac{ens(-vd\boldsymbol{l}) \times \boldsymbol{r}}{r^3} = \dfrac{\mu_0}{4\pi} \cdot \dfrac{-ensv}{r^3} d\boldsymbol{l} \times \boldsymbol{r}$$

所以

$$d\boldsymbol{B} = \dfrac{\mu_0}{4\pi} \cdot \dfrac{I d\boldsymbol{l} \times \boldsymbol{r}}{r^3} \tag{20}$$

运动电荷激发磁场过渡到电流元产生磁场的表达式就是式(20)，该式被称为微分形式的毕奥—萨伐尔定律。应用式(20)和磁感应强度叠加原理，对于整个长度为 l 的载流环路，矢量和的结果是在空间任意一点上各个电流元积分的结果

$$\boldsymbol{B} = \oint_{(l)} d\boldsymbol{B} = \dfrac{\mu_0}{4\pi} \oint_{(l)} \dfrac{I d\boldsymbol{l} \times \boldsymbol{r}}{r^3} \tag{21}$$

上面两个式子就被称为毕奥—萨伐尔定律的数学表达式。具体内容可以陈述为电流元 $Id\boldsymbol{l}$ 在空间任意一点 p 处激发的磁感应强度 $d\boldsymbol{B}$，其量值 dB 与 Idl 成正比，与 $\sin\theta$ 成正比，与 r^2 成反比；$d\boldsymbol{B}$ 的方向垂直于 $d\boldsymbol{l}$ 和 $\boldsymbol{r}$ 所决定的平面，指向为由 $Id\boldsymbol{l}$ 经小于 $180°$ 转向 $\boldsymbol{r}$ 的右手螺旋前进的方向[2]。

3　验证毕奥—萨伐尔定律的正确性

恒定磁场与静电场具有相似性，静电场是从库仑定律入手推导出来的，磁场中安培定律是与库仑定律形式相类似。其数学表达式是

$$d\boldsymbol{F} = \dfrac{\mu_0 I_2 d\boldsymbol{l}_2 \times (I_1 d\boldsymbol{l}_1 \times \boldsymbol{r}_{12})}{4\pi r_{12}^3}$$

其中 I_1, I_2 分别为两环路的电流，$d\boldsymbol{F}_{12}$ 是在两环路中分别选取的两电流元 $d\boldsymbol{l}_1$ 和 $d\boldsymbol{l}_2$ 之间的相互作用力，r_{12} 是两电流元之间的距离。与电场的情形相

---

[1] 励子伟,宋建平. 普通物理学·电磁学. 北京:北京大学出版社,1988:242-243.
[2] 洪佩智,韩树东. 工科大学物理学. 北京:北京理工大学出版社,1995:73.

比对 $I_2\mathrm{d}\boldsymbol{l}_2$ 看作试探电流元,将上式拆成两部分

$$\begin{cases} \mathrm{d}\boldsymbol{F}=I_2\mathrm{d}\boldsymbol{l}_2\times\mathrm{d}\boldsymbol{B} \\ \mathrm{d}\boldsymbol{B}=\dfrac{\mu_0}{4\pi}\cdot\dfrac{I_1\mathrm{d}\boldsymbol{l}_1\times\boldsymbol{r}_{12}}{r_{12}^3} \end{cases} \quad (22)$$

式(22)中的第一个式子称作磁感应强度的定义式,第二个式子是电流元所在处产生的磁感应强度公式,去掉下角标1,2有微分式

$$\mathrm{d}\boldsymbol{B}=\dfrac{\mu_0}{4\pi}\cdot\dfrac{I\mathrm{d}\boldsymbol{l}\times\boldsymbol{r}}{r^3} \quad (23)$$

积分公式为

$$B=\oint_{(l)}\mathrm{d}\boldsymbol{B}=\dfrac{\mu_0}{4\pi}\oint_{(l)}\dfrac{I\mathrm{d}\boldsymbol{l}\times\boldsymbol{r}}{r^3} \quad (24)$$

该过程看出,安培定律导出的和电磁场变换关系式导出的毕奥—萨伐尔定律的结果相同.那么,运动的电荷是引起一切磁现象的原因,电流产生的恒定磁场不过是导体中做定向运动的大量自由电子所激发的磁场的宏观表现.1911 年俄国物理学家约飞首先用实验证实,阴极射线管内的电子束同样在空间激发磁场,而且与具有等量的电流所激发的磁场一致.这进一步证明了导体中的运动电荷与空间中运动的带电粒子在激发磁场上是等效的[1].所以能够从导体内运动电荷与他所激发的磁场之间的定量关系出发,得到毕奥—萨伐尔定律.

本书的核心是定义电磁场矢量的思想实验,我国的许多优秀物理教师也都在做同样的工作.如东华大学理学院的浦天舒教授于 2015 年通过思想实验定义了电通量密度和磁场强度两个电磁场矢量,并认识其物理意义.通过高斯定律和安培定律并借助实验上的库仑定律和毕奥—萨伐尔定律,找出了电通量密度与电场强度,以及磁场强度与磁感应强度之间所必须满足的关系.

1 引 言

讲到电磁场,首先碰到的问题是场量的定义.在大多数大学物理教科书中通常只定义了电场强度 $\boldsymbol{E}$ 和磁感应强度(磁通量密度)$\boldsymbol{B}$,而把电通量密度(电位移)$\boldsymbol{D}$ 和磁场强度 $\boldsymbol{H}$ 作为辅助量引入.这样虽然在逻辑上没有问题,但 $\boldsymbol{D}$ 和 $\boldsymbol{H}$ 却没有明确的物理意义,而且容易给人造成 $\boldsymbol{E},\boldsymbol{B}$ 重

---

[1] 殷传宗,邓昭镜,罗琬华,等.基础物理专题选讲学.重庆:西南师范大学出版社,2006:174-175.

要,而 D, H 似乎不太重要的印象.但事实上我们知道电磁场是由4个矢量 E, B, D, H 构成的,它们应该是同等重要的(当然在不同的研究领域它们的重要性会有所不同).这一点从麦克斯韦方程组也可以看出.所以最好对这4个场量各自单独定义.这样既可赋予 D, H 一定的物理意义,而且借助实验定律也可以找出它们跟 E, B 的关系.

2 通过思想实验定义 D 和 H

4个电磁场矢量都可以通过所谓的"思想实验"进行定义[①].例如电场强度 E 和磁感应强度 B 可分别通过作用在静止和运动的试验电荷上的力的思想实验加以定义(借助法拉第定律,也可以从磁通量密度的概念通过思想实验来定义 B[②]).同样, D 和 H 也可以通过思想实验定义.

2.1 定义 D 的思想实验

D 一般称为电通量密度,但电通量本来是通过电场强度矢量 E 来定义的.即通过一个面元 dS 的电通量 $d\Phi_e$ 定义为

$$d\Phi_e = E \cdot dS \qquad (1)$$

因此我们也可以从电通量密度的概念来定义 E.但现在我们从矢量 D 来定义电通量

$$d\Psi_e = D \cdot dS \qquad (2)$$

可以通过如下思想实验来定义 D[③]:考虑将一对圆形金属电极板叠在一起放入由 D 形成的场中(图1(a)).假定极板的面积($\Delta S = \pi a^2$)很小,并且极板的厚度 δt 远小于极板的直径,即 $\delta t \ll 2a$.暂且假设, D 也是由电荷产生的,并且对电荷的作用跟 E 一样.则由于矢量场 D 的存在,极板中的电荷将发生位移,并在极板上形成表面电荷(积累负电荷的称为正极板,而积累正电荷的称为负极板).电荷位移后重排的结果使得在极板之间原来的场和由于表面电荷所产生的场相加的总场为零.这样就可以用极板间由表面电荷所产生的场来定义原来的场.为了做到这一点,现在将两块极板分开(图1(b)),发生了位移的电荷便留在了正负极板上.然后将极板从场中移出并接到一只电荷计(例如一只冲击电流计)上(图1(c)).于

---

① Paul R. Karmel, Gabriel D. Colef, Raymond L. Camisa. Introduction to Electromagnetic and Microwave Engineering. New York: John Wiley & Sons, Inc., 1998: 57-64.

② 同上.

③ 同上.

是先前留在极板上的电荷将通过电荷计给出一个总电荷读数(即冲击电流 $I(t)$ 之时间积分 $\int_0^t I(t)\mathrm{d}t$). 将此实验进行多次,每一次电极在场中取向不同. 在某个特定的取向,位移电荷最多,因此电荷计的读数会有一个最大值. 定义电位移矢量的大小等于产生的最大位移电荷量与极板的面积之比,方向为当极板取向使位移电荷量最大时正极板的法线方向. 在数学上,如果用 D 代表电位移矢量, a_n 代表产生最大位移电荷量 ΔQ_0 时正极板到负极板的单位法线矢量,那么

$$D = a_n \left[\lim_{\Delta S \to 0} \frac{\Delta Q_0}{\Delta S} \right]_{\max} \tag{3}$$

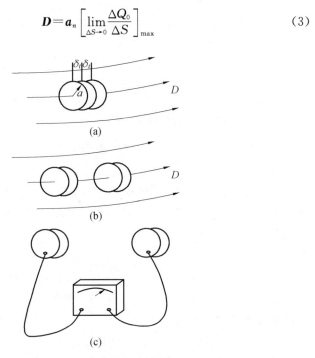

图1 矢量场 D 中的一对薄圆形导电金属电极

D 叫作电通量密度是因为它给出了通过单位面积的场的力线的数目(称为通量),而符号"D"则表示它是由电荷的位移(Displacement)产生的场,这一名称是与上述思想实验中导电极板上的位移电荷相联系的.

从 D 的定义可以引出高斯定律. 因为对一个闭合面来讲, D 在整个闭合面的外法线分量上的积分便是 ΔQ_0,所以 ΔQ_0 也就是抵消原来场的位移电荷,所谓高斯定律便是

$$\Psi_e = \oiint D \cdot \mathrm{d}S = \Delta Q_0 \tag{4}$$

即漏出一个闭合面的电通量等于由这个闭合面所包围的总电荷.

现在我们假设 ΔQ_0 为一点电荷，则由于球对称性，在距离 ΔQ_0 为 R 的球面上任一点处有 $\boldsymbol{D}=\boldsymbol{a}_R D(R)$，由高斯定律可以得到

$$\boldsymbol{D}=\boldsymbol{a}_R D(R)=\frac{\Delta Q_0}{4\pi R^2}\boldsymbol{a}_R \tag{5}$$

若我们在距离 ΔQ_0 为 R 的球面上任一点放置一静止测试电荷 Q_t，则如果 $\boldsymbol{D}=\varepsilon_0\boldsymbol{E}$，那么这个测试电荷所受的力为

$$\boldsymbol{F}=Q_t\boldsymbol{E}=\frac{1}{4\pi\varepsilon_0}\frac{Q_t\Delta Q_0}{R^2}\boldsymbol{a}_R \tag{6}$$

这正是库仑定律，即如果我们假定高斯定律中的 ΔQ_0 是自由电荷，那么 $\boldsymbol{D}=\varepsilon_0\boldsymbol{E}$ 便是自由空间中 $\boldsymbol{D}$ 与 $\boldsymbol{E}$ 必须满足的关系。

当存在媒质时，媒质分子会因极化而产生极化电荷，与自由电荷不同的，极化电荷是正负电荷错开形成的分子偶极矩，所以若因"位移"而穿出闭合面的极化电荷是 $-\Delta Q'$ 的话，则在闭合面内便有剩余电荷 $\Delta Q'$。对于极化电荷，我们定义极化矢量 $\boldsymbol{P}$

$$\boldsymbol{P}=\boldsymbol{a}_n\left[\lim_{\Delta S\to 0}\frac{-\Delta Q'}{\Delta S}\right]_{max} \tag{7}$$

则 $\boldsymbol{P}$ 在整个闭合面的外法线分量上的积分

$$\oiint \boldsymbol{P}\cdot\mathrm{d}\boldsymbol{S}=-\Delta Q' \tag{8}$$

可以证明，此定义与一般书上以单位体积内微观分子电偶极矩的矢量和定义的 $\boldsymbol{P}$ 是一致的[①]。这样等于是把 $\boldsymbol{D}$ 看成是自由电荷产生的场（不论是在媒质中还是在自由空间），把 $\boldsymbol{P}$ 看成是极化电荷产生的场，只是极化电荷产生的场并不能像自由电荷产生的场那样抵消原来的场（即自由电荷的场），所以 $\boldsymbol{P}$ 和 $\boldsymbol{D}$ 是不同的场矢量。但不论是在媒质中还是在自由空间，$\boldsymbol{D}$ 都由式(3)定义。

可令媒质中的总电荷 $Q=\Delta Q_0+\Delta Q'$，这里 ΔQ_0 代表自由电荷。于是由式(4)和式(8)有

$$\oiint(\boldsymbol{D}-\boldsymbol{P})\cdot\mathrm{d}\boldsymbol{S}=\Delta Q_0+\Delta Q'=Q \tag{9}$$

即总电荷产生的场是 $\boldsymbol{D}-\boldsymbol{P}$。上式也可看成是矢量 $\boldsymbol{D}-\boldsymbol{P}$ 的高斯定律。如果 Q 为一点电荷，则同样有

$$\boldsymbol{D}-\boldsymbol{P}=\frac{Q}{4\pi R^2}\boldsymbol{a}_R \tag{10}$$

若

---

① 赵凯华，陈熙谋. 电磁学(上册). 北京：人民教育出版社，1978：141-143，290.

$$\boldsymbol{D}-\boldsymbol{P}=\varepsilon_0\boldsymbol{E} \tag{11}$$

则位于 R 处的测试电荷 Q_t 所受的力

$$\boldsymbol{F}=Q_t\boldsymbol{E}=\frac{1}{4\pi\varepsilon_0}\frac{Q_tQ}{R^2}\boldsymbol{a}_R \tag{12}$$

满足库仑定律.所以式(11)就是媒质中 $\boldsymbol{D}$ 与 $\boldsymbol{E}$ 必须满足的关系.因此我们可以把式(11)看成是由高斯定律借助库仑定律得到的.也可以把库仑定律看成是由高斯定律借助式(11)得到的.

从上述思想实验我们看到,$\boldsymbol{D}$ 可以看成是自由电荷产生的场,而真正的电场 $\boldsymbol{E}$ 则包括了所有电荷的贡献,但 $\boldsymbol{D}$ 和 $\boldsymbol{E}$ 都对所有电荷(自由电荷与极化电荷)有作用.

2.2 定义 H 的思想实验

$\boldsymbol{H}$ 可以通过如下思想实验定义:取一用导线密绕成一个圆柱形状的螺线管线圈(图 2(a)).螺线管的长度 ΔL 很短,但与它截面的直径 $2a$ 相比仍很长,即 $\Delta L \gg 2a$,其上绕有 n 圈导线.连接一电流源至线圈导线,便会在导线中产生电流 I.螺线管的单位长度上的电流因而是 $\frac{nI}{\Delta L}$.可把此螺线管理想化地看成是一个围绕着一薄层表面电流 $J_{s0}=\frac{nI}{\Delta L}$ 的短圆柱体,如图 2(b)所示.

现在,把一棒形磁体放入螺线管,并测量这一磁体上受到的力矩.结果发现力矩的强度在螺线管内是均匀的(只要放置的磁体远离两端),并且正比于电流 I,而此力矩的方向是使磁棒沿螺线管的轴线.我们定义螺线管内的磁场强度的大小等于表面电流 J_{S0},方向为当右手四指朝电流方向握住螺线管时大拇指所指的螺线管的轴向,其单位矢量为 $\boldsymbol{a}_z$(图 2(c)),即

$$\boldsymbol{H}=J_{S0}\boldsymbol{a}_z \tag{13}$$

为了测量空间一点 P 的磁场强度,我们可以把这个通电小螺线管放在这一点,然后调节电流的大小以及螺线管的取向,直到使螺线管内小磁棒上的力矩为零,那么螺线管内部的磁场一定为零.由于现在的总场是我们要测量的原始场和螺线管电流产生的场之和,所以原始场的大小等于螺线管电流产生的场,但方向相反.这一测量也可以在材料媒质中进行,只要在我们的思想实验中想象这个小螺线管被固定在材料媒质中的一个很细的管道中.

如果我们把 $\boldsymbol{H}$ 定义为仅由自由电荷的流动形成的电流 I_0(来自导电媒质或自由空间中移动的电子、质子或离子的流动)产生的场,那么在磁

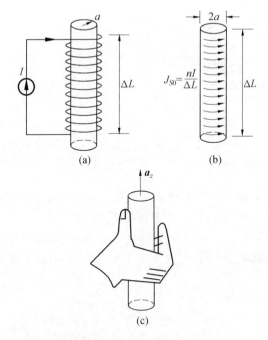

图2 定义磁场强度的密绕通电螺线管线圈

化了的媒质中,可以设想一个由表面磁化电流 nI' 形成的螺线管,定义一个磁化强度矢量 M

$$M = \frac{nI'}{\Delta L} a_z \tag{14}$$

式中 a_z 的方向也是按右手规则确定的. 注意 M 并不能使磁场 H 中的螺线管内小磁棒上的力矩为零. 因为单匝线圈的磁化电流 I' 产生的磁矩为 $m = I' \Delta S a_z (\Delta S = \pi a^2)$,所以式(14)也表示体积元 $\Delta L \Delta S$ 内的磁矩之和,即 $M = \frac{nm}{\Delta L \Delta S} = \frac{nI'}{\Delta L} a_z$,故有[①]

$$\oint M \cdot dl = nI' \tag{15}$$

而对于 H 显然应该相应的有

$$\oint H \cdot dl = nI_0 \tag{16}$$

式中 I_0 为自由电荷的流动产生的电流. 这就是安培定律.

因为由实验上的毕奥—萨伐尔定律可以导出单位长度上螺线管内的

---

① 赵凯华,陈熙谋. 电磁学(下册). 北京:人民教育出版社,1978:85-88.

磁感应强度为[①]

$$B = \mu_0 \frac{nI'}{\Delta L} a_z \tag{17}$$

其方向 a_z 也由右手规则确定,但这里的电流包括了磁化电流的总电流,即 $I = I_0 + I'$.因此磁感应强度与电流的积分关系可写成

$$\oint B \cdot dl = \mu_0 n(I_0 + I') \tag{18}$$

由式(15),(16)以及式(18),有

$$\frac{B}{\mu_0} = H + M \tag{19}$$

在一般电磁学书上安培定律是由毕奥—萨伐尔定律导出的,而在这里我们通过思想实验定义了 H,相当于先假定安培定律,但为了使由毕奥—萨伐尔定律得到的结果式(17)成立,H 与 M 必须满足式(19).当然我们也可以把毕奥—萨伐尔定律看作是由安培定律借助式(19)而得到的.

3 结　论

我们在非时变的静电和静磁情况下通过思想实验定义了电磁场矢量 D 和 H.在第一个思想实验中,电荷位移使电极之间的场为零实际上是假定高斯定律成立;在第二个思想实验中,使放入场中的螺线管中的小磁棒受到的力矩为零实际上是假定安培定律成立.而式(11)和式(19)必须在实验定律即库仑定律和毕奥—萨伐尔定律成立的条件下得到,或者说借助式(11)和式(19)可由高斯定律和安培定律导出库仑定律和毕奥—萨伐尔定律.在一般电磁学书上通常是由库仑定律和毕奥—萨伐尔定律这两个实验定律导出高斯定律和安培定律的,但我们知道库仑定律原先是在静电场中得到的实验定律,而高斯定律则被麦克斯韦推广到了时变的情况(这样等于也把库仑定律推广到了动态情形);同样,在麦克斯韦方程组中安培定律也是一个基本定律(不过在把它推广到时变情况时加上了位移电流项).所以高斯定律和安培定律(时变时应加上位移电流项)是更普遍的定律,由高斯定律和安培定律导出库仑定律和毕奥—萨伐尔定律在逻辑上更合理.

本书的内容与我们的生活息息相关,作为一个现代人不可不知.如果还记得一

[①] 赵凯华,陈熙谋.电磁学(上册).北京:人民教育出版社,1978:141-143,290.

点中学物理知识的话,会知道地球自成形以来就浸没在电磁波的海洋中,一直到今天还是这样.可见光、红外光、紫外光都是电磁波.如果从来没有电磁波,就一直不会有生命;如果现在电磁波统统消失,人类会在不久以后死于饥寒交迫.我们世世代代都浸没在电磁波的海洋里:红外线温暖我们,紫外线伤害我们,可见光令我们的视觉细胞享受色彩的盛宴.在雨后湿润的空气里,白色的阳光散射成彩虹,渺小的人类得以窥视一点电磁波的宏大奥秘.凭着这一点现象,人类中的杰出者写出了精确描述电磁波的麦克斯韦方程."阳春布德泽,万物生光辉."说的就是电磁波的德泽和光辉.

刘培杰

2021 年 3 月 31 日

于哈工大

刘培杰物理工作室
已出版(即将出版)图书目录

序号	书　名	出版时间	定　价
1	物理学中的几何方法	2017—06	88.00
2	量子力学原理.上	2016—01	38.00
3	时标动力学方程的指数型二分性与周期解	2016—04	48.00
4	重刚体绕不动点运动方程的积分法	2016—05	68.00
5	水轮机水力稳定性	2016—05	48.00
6	Lévy噪音驱动的传染病模型的动力学行为	2016—05	48.00
7	铣加工动力学系统稳定性研究的数学方法	2016—11	28.00
8	粒子图像测速仪实用指南:第二版	2017—08	78.00
9	锥形波入射粗糙表面反散射问题理论与算法	2018—03	68.00
10	混沌动力学:分形、平铺、代换	2019—09	48.00
11	从开普勒到阿诺德——三体问题的历史	2014—05	298.00
12	数学物理大百科全书.第1卷	2016—01	418.00
13	数学物理大百科全书.第2卷	2016—01	408.00
14	数学物理大百科全书.第3卷	2016—01	396.00
15	数学物理大百科全书.第4卷	2016—01	408.00
16	数学物理大百科全书.第5卷	2016—01	368.00
17	量子机器学习中数据挖掘的量子计算方法	2016—01	98.00
18	量子物理的非常规方法	2016—01	118.00
19	运输过程的统一非局部理论:广义波尔兹曼物理动力学,第2版	2016—01	198.00
20	量子力学与经典力学之间的联系在原子、分子及电动力学系统建模中的应用	2016—01	58.00
21	动力系统与统计力学:英文	2018—09	118.00
22	表示论与动力系统:英文	2018—09	118.00
23	工程师与科学家微分方程用书:第4版	2019—07	58.00
24	工程师与科学家统计学:第4版	2019—06	58.00
25	通往天文学的途径:第5版	2019—05	58.00
26	量子世界中的蝴蝶:最迷人的量子分形故事	2020—06	118.00
27	走进量子力学	2020—06	118.00
28	计算物理学概论	2020—06	48.00
29	物质,空间和时间的理论:量子理论	2020—10	48.00
30	物质,空间和时间的理论:经典理论	2020—10	48.00
31	量子场理论:解释世界的神秘背景	2020—07	38.00
32	计算物理学概论	2020—06	48.00
33	行星状星云	2020—10	38.00

刘培杰物理工作室
已出版(即将出版)图书目录

序号	书　名	出版时间	定价
34	基本宇宙学:从亚里士多德的宇宙到大爆炸	2020—08	58.00
35	数学磁流体力学	2020—07	58.00
36	高考物理解题金典(第2版)	2019—05	68.00
37	高考物理压轴题全解	2017—04	48.00
38	高中物理经典问题25讲	2017—05	28.00
39	高中物理教学讲义	2018—01	48.00
40	1000个国外中学物理好题	2012—04	48.00
41	数学解题中的物理方法	2011—06	28.00
42	力学在几何中的一些应用	2013—01	38.00
43	物理奥林匹克竞赛大题典——力学卷	2014—11	48.00
44	物理奥林匹克竞赛大题典——热学卷	2014—04	28.00
45	物理奥林匹克竞赛大题典——电磁学卷	2015—07	48.00
46	物理奥林匹克竞赛大题典——光学与近代物理卷	2014—06	28.00
47	电磁理论	2020—08	48.00
48	连续介质力学中的非线性问题	2020—09	78.00
49	力学若干基本问题的发展概论	2020—11	48.00
50	狭义相对论与广义相对论:时空与引力导论(英文)	2021—07	88.00
51	束流物理学和粒子加速器的实践介绍:第2版(英文)	2021—07	88.00
52	凝聚态物理中的拓扑和微分几何简介(英文)	2021—05	88.00
53	广义相对论:黑洞、引力波和宇宙学介绍(英文)	2021—06	68.00
54	现代分析电磁均质化(英文)	2021—06	68.00
55	为科学家提供的基本流体动力学(英文)	2021—06	88.00
56	视觉天文学:理解夜空的指南(英文)	2021—06	68.00
57	物理学中的计算方法(英文)	2021—06	68.00
58	单星的结构与演化:导论(英文)	2021—06	108.00
59	超越居里:1903年至1963年物理界四位女性及其著名发现(英文)	2021—06	68.00
60	范德瓦尔斯流体热力学的进展(英文)	2021—06	68.00
61	先进的托卡马克稳定性理论(英文)	2021—06	88.00
62	经典场论导论:基本相互作用的过程(英文)	2021—07	88.00
63	光致电离量子动力学方法原理(英文)	2021—07	108.00
64	经典域论和应力:能量张量(英文)	2021—05	88.00
65	非线性太赫兹光谱的概念与应用(英文)	2021—06	68.00
66	电磁学中的无穷空间并矢格林函数(英文)	2021—06	88.00
67	物理科学基础数学.第1卷,齐次边值问题、傅里叶方法和特殊函数(英文)	2021—07	108.00
68	离散量子力学(英文)	2021—07	68.00

刘培杰物理工作室
已出版(即将出版)图书目录

序号	书 名	出版时间	定 价
69	核磁共振的物理学和数学(英文)	2021—07	108.00
70	分子水平的静电学(英文)	2021—08	68.00
71	非线性波:理论、计算机模拟、实验(英文)	2021—06	108.00
72	石墨烯光学:经典问题的电解解决方案(英文)	2021—06	68.00
73	超材料多元宇宙(英文)	2021—07	68.00
74	银河系外的天体物理学(英文)	2021—07	68.00
75	原子物理学(英文)	2021—07	68.00
76	将光打结:将拓扑学应用于光学(英文)	2021—07	68.00
77	电磁学:问题与解法(英文)	2021—07	88.00
78	海浪的原理:介绍量子力学的技巧与应用(英文)	2021—07	108.00
79	杰弗里·英格拉姆·泰勒科学论文集:第1卷.固体力学(英文)	2021—05	78.00
80	杰弗里·英格拉姆·泰勒科学论文集:第2卷.气象学、海洋学和湍流(英文)	2021—05	68.00
81	杰弗里·英格拉姆·泰勒科学论文集:第3卷.空气动力学以及落弹数和爆炸的力学(英文)	2021—05	68.00
82	杰弗里·英格拉姆·泰勒科学论文集:第4卷.有关流体力学(英文)	2021—05	58.00
83	多孔介质中的流体:输运与相变(英文)	2021—07	68.00
84	洛伦兹群的物理学	2021—08	68.00
85	物理导论的数学方法和解决方法手册	2021—08	68.00

联系地址:哈尔滨市南岗区复华四道街10号　哈尔滨工业大学出版社刘培杰物理工作室
网　　址:http://lpj.hit.edu.cn/
邮　　编:150006
联系电话:0451—86281378　　13904613167
E-mail:lpj1378@163.com